I0037369

THE
BASIC
SCHOOL

The Making of Marine Officers
at Quantico, Virginia

Thomas Smith

Chapbook Press

Schuler Books
2660 28th Street SE
Grand Rapids, MI 49512
(616) 942-7330
www.schulerbooks.com

The Basic School: The Making of Marine Officers at Quantico, Virginia

ISBN 13: 9781957169385

Library of Congress Control Number: 2023903929

Copyright © 2023 Thomas Smith

All rights reserved. No part of this publication may be reproduced, stored in a retrieval system, or transmitted in any form by any means—electronic, mechanical, photocopying, recording, or otherwise—except for the purpose of brief reviews, without written permission of the author.

Printed in the United States by Chapbook Press.

To all Marine officers who have gone through the gates of Quantico and experienced the fear, the loneliness, the pain, the *esprit de corps,* and the euphoria of graduating TBS.

The Marines have a school that reflects a concept the other services would do well to copy: The Basic School.

> Former U.S. Senator and former Chairman of the Armed Services Subcommittee, Gary Hart, in his book *America Can Win.*

Lieutenants of Fox Company, 6-91, you are graduating from the only service school of its kind in the American military establishment and the world for that matter. Regardless of future occupational specialty, you have all been trained in the rudiments of being rifle platoon commanders.

> Major P.B. Johnson, Company Commander, Fox Company, TBS Class 6-91

In every war, Marines have borne a heavy burden far out of proportion to their limited numbers. This fact stands as a silent tribute to the individual Marine. Our present Basic School student will soon be leading Marines. It is your mission to educate this young officer and thereby make him worthy of this privilege.

> Inscription over the passageway leading from Heywood Hall to the classrooms at The Basic School

TBS is four years of college crammed into six months.

Anonymous

A pint of sweat in training saves a gallon of blood in war.

General George S. Patton, Jr.

Men are neither lions nor sheep. It is the man who leads them who turns them into either lions or sheep.

Jean Dutourd, *The Taxis of the Marne*

The safest place in Korea was right behind a platoon of Marines. Lord, how they could fight.

Army Major General Frank E. Lowe

Marine human material was not one whit better than that of the human society from which it came. But it had been hammered into form in a different forge.

T.R. Fehrenbach, *This Kind of War*

Contents

The Etymology of TBS

Marines refer to "The Basic School" by its initialism "TBS." However, "TBS" stands for much more than just "The Basic School." Here are some irreverent, not to mention insightful, ways that enterprising Marine lieutenants have defined "TBS":

The Big Suck
The Big Suffering
Testicles Being Slapped
The Bleeding Sphincter
The Butterbar School
Trainees Being Saluted
This Base Sucks
These Barracks Smell
The Back Stab
The Body Softener
Tired Back Syndrome
The Broken System
The Bleakest Season
Tension Building School
Tactical Boy Scouts
Tactics Being Slaughtered
Totally Broken Spirits
Thousands of Boring Slides
Thousands of Blank Stares
Thousands Bored Senseless
Text Book Solutions
The Box Search
The Bummer Summer

Taxes Being Squandered
Taxpayers Biggest Sinkhole
Ticks, Bugs, Spiders
Ticks, Beavers, Snakes
Termination Bachelor Status
Tewts, Boxes, Stexes
Too Bad Sucker
Try Being Secured
Toddlers Being Salty
Temporarily Behind Schedule
Totally Bogus Scoop
Total B.S.
Time Between Saturdays
Tortured By Sadists
The Big Shuffle
Totally Biting Sarcasm
The Bubba School
Time Being Squandered
Time Being Sodomized
The Best Scam
Tedious But Satisfying
The Bridal School
Thorns, Briars, Stickers
The Bogus Scenario
Trivia Becoming Sacred
Terminal By Sundown
Tired Before Sunrise
TS&L, BAMCIS, SMEAC

5-Paragraph Order

U.S. Marines are taught a concept known as the 5-Paragraph Order.[3] It's an order issued by Marines to squads, platoons and companies. Accordingly, it's only appropriate that the reader, before reading this book about The Basic School and the making of Marine officers, be given a 5-Paragraph Order. So standby for the 5-Paragraph Order and hold all questions to the end:

1. *Situation*—**You must understand who this book is written for.** This book is written principally for those who want to become Marine officers. It is also written for those who have been through the gates of Quantico and experienced Marine Corps officer training. Finally, this book is written for anyone who wants to lift the veil and see what Marine training is like and how it's done.

2. *Mission*—**You must understand why this book was written and what it's about.** I wrote this book for two reasons: The first is because of the importance and lasting impact that TBS has had on Marine officers through the years. Many Marine officers will praise TBS and its training program long after graduation and well into their

---

1   Marines use an initialism for the 5-Paragraph Order: "**SMEAC**": (1) *Situation* (enemy strength, disposition, capabilities, priorities); (2) *Mission* (who, what, when, where, why); (3) *Execution* (attack plan, operating instructions, intent, specific assignments); (4) *Admin & Logistics* (bullets, bandages, beans, bad guys); and (5) *Command and Control* (leader location, contingency signals, emergency signals).

careers in the Big Green Machine. The second reason I wrote the book is because of the paucity of information on the making of Marine officers, and specifically TBS. To my knowledge, no book or monograph on TBS exists.

3. _Execution_—You must understand that this book does not purport to reflect the current curriculum at TBS. Nor can any book. The curriculum at TBS is constantly changing to reflect real-world scenarios. In fact, the TBS curriculum changed many times _during_ my class, and immediately after it, as well. Therefore, one thing is certain: The TBS curriculum in place today is not what it was yesterday. Nor will tomorrow's curriculum be the same as today's. So writing a completely accurate, contemporary account of TBS's curriculum is impossible. But that doesn't mean a book or article should not be written about TBS. Very much remains the same, much more than is different. After all, there must be enough similarity and consistency to the curriculum and standards to allow a Marine to graduate and be called a "Marine officer." There must be enough consistency to have the respect of his peers and other Marines—past, present, and future—along with citizens and national leaders. For the unconvinced, here are some timeless verities experienced by Marine lieutenants going through TBS, regardless of the curriculum: The immersion in the Marine

Corps culture and its cauldron of values, exacting standards, and discipline; the lack of sleep; the physical exhaustion; the fear of physical injury; the effects of the cruel, unharnessed weather; the comraderie between lieutenants; and the extreme euphoria of tasting the nectar and graduating TBS to become bonafide officers of Marines.

4. *Admin & Logistics*—You must understand that this book is a work of nonfiction. All events and stories depicted in this book are true. However, the events and stories are only as accurate as the author's notes and memory. This book was mainly written over the course of a decade, some parts reduced to writing only minutes and hours after the events, some weeks after, some months and years after. All factual inaccuracies and omissions are, however, the sole responsibility of the author.

5. *Command & Control*—You must understand that this book is my view of the elephant. Nothing more, nothing less. Approximately 230 hard-charging Marine lieutenants attended TBS Class 6-91, from September 23, 1991 to April 2, 1992, and I cannot, and do not, profess to speak for all of them. It's like the old proverb of the blind men feeling the elephant. They each have their own way of feeling, processing, and coming to conclusions about the beast. The same holds true with TBS. While TBS itself is factual, external, and seemingly one huge thing, with an unbelievably difficult curriculum and a start- and end-date, it is not the same huge thing to everyone. It is many things at once, depending on one's vantage point. As a result, I'm sure some current and former Marine officers may quarrel or take umbrage with my interpretation of some events. My response? They may be right.

What is TBS?

The Basic School (TBS) is the 2nd school in the progression of making Marine officers. It's located at Marine Corps Base-Quantico, Virginia. The 1st school is Officer Candidate School (OCS), also located at Quantico; the 3rd school is the Military Occupational School (MOS), located at various Marine bases or Army Posts across the nation. OCS is the screening and selection school, winnowing out the physically and mentally weak through extremely difficult physical and mental events. The MOS school is the training school for chosen occupations, such as infantry, artillery, communications, combat engineering, armor, naval justice, and so on. Sandwiched between OCS and the MOS schools is TBS. TBS is where Marine lieutenants learn how to be Marine officers. Nothing more, nothing less. Here's TBS's mission statement:

> Train and educate newly commissioned or appointed officers in the high standards of professional knowledge, *esprit-de-corps,* and leadership required to prepare them for duty as company grade officers in the operating forces, with particular emphasis on the duties, responsibilities and war fighting skills required of a rifle platoon commander.

Specifically, this one-sentence paragraph boils down to two themes. The first is that TBS is a "socialization" process, where it teaches its officers about the Marine Corps and what the Marine Corps requires of them: learning its history, traditions, *esprit de corps*, and imbuing into them values, such as integrity, honesty, truthfulness, hard work, and professionalism. The second theme is tactics and the rudiments of war. In short, TBS transforms Marine officers and turns them into a force of competent and cohesive warriors. As a result, TBS's importance as an institution in the Marine Corps is only equaled by its two recruit depots in San Diego and Parris Island.

My TBS class ran from Monday, September 23, 1991 to Thursday, April 2, 1992, spanning 25 weeks, or roughly six months, excluding three weeks of leave used over Thanksgiving and Christmas holidays. It had 230 Marines in it. It totaled over 1,500 hours of instruction, which approximates what American universities require their students to complete over four years. In other words, TBS is a school that crams four years of college into six months.

A Historical Perspective

Trying to compare TBS with other military schools across the globe is a difficult endeavor. The schools are all different, with different mission statements, characteristics, cultures, and curricula. But one school stands out, at least to this author, when comparing it to TBS: the *Kriegsakademie*, a school formed by the vaunted German Army in the 19th Century. Operating from 1815 to 1945, the *Kriegsakademie* was one of the world's great military schools. It was dedicated solely to the study of war and the training of officers. Its three-year curriculum focused on the rudiments of war: strategy, tactics, weaponry, military history, and fortifications. The *Kriegsakademie* had an unparalleled academic reputation, both inside and outside the military, as good as the best civilian universities today. But the worth of the *Kriegsakademie* was not so much in its academic prowess as in its ability to train officers for war, or, in the words of Trevor Dupuy, to "institutionalize military excellence." These officers led the German Army from one victory to another. Martin Van Creveld, in his book *Training for Officers*, noted that "to judge by results achieved between 1866 and 1945, the German system for teaching officers how to command in war has never been equaled in the modern world." What this shows is that the *Kriegsakademie* was not limited to improving the competence of a single professional; it even transformed the officer corps into a force of competent warriors. On a smaller scale, this is what TBS does. It institutionalizes military excellence. It transforms Marine officers into a force of competent warriors. Certainly, TBS is no *Kriegsakademie*, but it does transform its officers, and it does capably train them in the rudiments of war. TBS is only six months' long, compared to the *Kriegsakademie's* three years, but TBS has over 1,500 hours of instruction, which is tantamount to four years of college.

Part *Kriegsakademie,* part War College, part university, part seminar, and part OJT, TBS is the oldest Marine Corps' school, established on May 1, 1891. Originally named the "School of Application," it was founded by Colonel Charles Heywood, the ninth commandant of the Marine Corps, and located at Washington, D.C., at the intersection of "8th and I" streets—which is the current home of the Commandant of the Marine Corps and its "Silent Drill Platoon." Through the years, the name eventually changed to "The Basic School," or "TBS," and moved from Washington, D.C. to Annapolis, Maryland to Norfolk, Virginia to Port Royal, South Carolina (now Parris Island) to the Philadelphia Naval Yard to, ultimately, its current resting place at Camp Barrett, Quantico, Virginia. The duration of TBS has also changed through the years (as an example, it was 15 weeks long in 1925 when General Lewis B. "Chesty" Puller attended; 26 weeks in 1966-1967; 21 weeks in 1968-1972; 26 weeks in 1973-1976; and 21 weeks in 1977-1987).[4]

---

4 *Marine Officer's Guide (4th Edition),* p. 262.

PROLOGUE

The Clones

There I was, seated in Classroom 1 of Heywood Hall, in a sea of bald heads, all of which were connected to strong jaw lines, exquisitely starched camouflage utilities, and black combat boots. But these bald heads were not entirely bald: many had small divots of hair on top, like a wide but very short Mohawk, with an electric razor swath mowed right down the center from front to rear—in effect, if you thought about it, a reverse Mohawk. Everyone in the classroom was young, about 22, extraordinarily fit, all male, and all adorned in the uniform-of-the-day, which turned out to be the uniform-of-everyday, with minor exceptions: starched and pressed camouflage utilities—known as cammies—with lieutenant's bars on the collars, and shiny black combat boots, known as Black Cadillacs. We were Marine clones assembled like seated rows of corn for classroom instruction. It was 0700 hrs on Monday, September 23, 1991, the first day of a very long and grueling six-month program called The Basic School (TBS) at the Marine Corps Base in Quantico, Virginia. The 100-hour Gulf War had ended earlier in the year. Anita Hill was rocking America in the Clarence Thomas confirmation hearings. I was rocking Quantico in cammies and Black Cadillacs, wearing lieutenant's bars.

There were 230 of us in this huge classroom, and, oh no, we weren't arranged haphazardly. We were arranged according to Marine Corps' procedure, with each of the

1

five equally-divided platoons having their own section of the classroom—all to preserve a concept known as "platoon integrity." I was ensconced in the cookie-cutter section of 5th platoon, looking terribly similar to everyone around me. The classroom, which was filled with testosterone up to here, was one noisy place. Student billet holders—Fire-Team Leaders, Squad Leaders, Platoon Commanders—were standing in the aisles yelling at everyone to hurry up and take a seat. *"Everyone, sit down!…I can't see!…Fire-team leaders, I need your counts!…We're "up!"…We're missing Suggs!… No, here he is, walking in…All present!…We're missing one Squad Leader!…Zeman!…he's at sick-call!…"* This was the process, repeated many times a day, for getting "counts" on who was present and who was UA (Unauthorized Absence, the contemporary term of art for AWOL, Absent WithOut Leave, which no longer is used in the U.S. military). But finally, after a short while, the movement and jiggling and positioning settled, the cacophony of sounds reduced to near silence with 230 bald heads clasping their hands on their desk-tops, staring straight ahead, waiting. It was church quiet. We were waiting for a "Welcome Aboard" presentation by Major Preston J. Johnson, the Commanding Officer (CO) of Fox Company, *our* company. He was at the front of the classroom behind a lectern, staring hard at us minions. He was our new Commander, Viceroy, Potentate, Icon, War Chief…our Yahweh for the next six months…He was Da Man!

Our *Student* Company Commander, who was the unluckiest man of the moment—he was the first lieutenant to get tagged at TBS with a company-level billet that carried the most responsibility—stood centered, facing us, at the front of the classroom. *Too Bad Sucker.*[3] He was waiting for the counts from each of the five Student Platoon Commanders. Once he received a "thumbs-up" from each of the five Student Platoon Commanders, he stiffened his body to the position of attention and croaked, "FOX COMPANY!...A-TEN-HOOOOOTT!" All at once 230 lieutenants kicked back their chairs and stood at attention, laser straight, feet together at the heels, toes at 45-degree angles, arms down by their sides, thumbs aligned along the trouser seams, staring straight ahead, all silent. The Student Company Commander executed a sharp about-face maneuver. Major Johnson slowly moved in front of him and stared into his face two feet away. The Student Company Commander, quietly, in almost a whisper, told Major Johnson how many Marines were on deck and prepared for instruction. The Major quietly answered, "Very well. Carry on." The student said, "Aye-Aye, Sir," executed a near flawless about-face, and screamed a guttural "FOX COMPANY!...TAKE YOUR SEATS!" This was the procedure for every class, every work day, until the end of TBS (you could, of course, insert Instructor-of-the-Hour for Major Johnson, who did not usually attend our classes).

Major Johnson resumed his place behind the lectern. We stared at him hard, sizing him up. He was 36, balding, five-eleven, fit, wore glasses and old, faded camouflage utilities—a mark of someone known as "salty," someone who'd been around awhile. Overall, he was unprepossessing in appearance, not anything like the motivated, nail-eating

---

3 Interspersed throughout this book, when relevant, are alternative names for "TBS" and "The Basic School" that insightful lieutenants have created over the years.

type I had expected. He was, in a word, "normal." As we came to find out, he was anything but normal; he was an amalgam of everything you'd want in a leader of Marines. He was articulate, exceedingly bright, knowledgeable, compassionate, competent. He was a leader who came to love his company—and his company, in turn, came to love him. And revere him. Consider just one small thing he did. He memorized all our names and significant details of our backgrounds. 230 bald Lieutenants! Think about that. He would stand in front of 230 Marines, 230 near clones, and if a lieutenant raised his hand, he would confidently call the lieutenant by name. At first, we thought it a fluke. Like he knew the lieutenant, or just met him. Then he would accurately call names, one after another, day after day, week after week, month after month. *In a sea of bald heads! Clones! How could he tell the difference? It's magic!* We couldn't believe it. To experience this in real-time and in real life was shocking. And it didn't just happen in the classrooms. When Major Johnson saw you out in the field, even before he met you, he would approach and tell you—out of the blue!—about an experience he had in your hometown or something he had heard about your home town. How did he know what town? How did he recognize me? We were usually wearing Kevlar helmets or soft covers, making it even more difficult to ID someone. After a month or two, we did what new lieutenants do best: we asked questions. In fact, we asked him. "Before the company started up, I would pull your files, five at a time, stare at your photos, and memorize your names and backgrounds until I knew everyone's," he said. "It's that simple." Simple? Yeah, right. Two hundred and thirty files—*230 bald heads!*—five at a time, which amounted to 46 days of memorizing. That was the beginning of our reverence for Major Johnson.

Major Johnson looked at us, hard. We looked back at him, just as hard. Silence filled the large room. He stepped from behind the lectern.

"Good morning Fox Company!" he said.

"**GOOD MORNING, SIR**!" we responded in unison, raising the roof.

"Welcome to The Basic School! I am Major Preston Johnson, your Company Commander for the next six months."

He returned to the lectern and studied it. He told us about his background, where he was raised, and about his 15-year career with the Marines. He launched into telling us the essentials, the rules and regulations—the coal that fires the Marine Corps. He told us how we would dress, act, and speak, on- and off-duty. He distributed a written "Welcome Aboard" packet that contained, of course, everything he was telling us. But he went on to add, extemporaneously, what we were to expect from him and TBS and its instructors. He said the training schedule would be nearly as filled and hectic as the one at OCS, and that we would, once again, be tested physically, mentally, and emotionally. But TBS, he said, would not be as physically crushing as OCS. *Amen, brother!* But he did add that there would be little time on the training schedule for physical training, even though there were major PT Events which were graded and weighted. He said we wouldn't have enough time to stay in shape during the training schedule. Everyone had to stay in shape for the PT events after the "close of business."

He congratulated us for graduating OCS. He was convinced that we were, because we graduated Marine OCS, the best of the best, the cream which rises to the top. He said he was confident that we could, as a whole, compete with any group of 230 individuals across the U.S. in academics, physical fitness, or in leading men—which, after all, was the chief purpose of Marine officer training. At this point, if Major Johnson was a public corporation, we would have bought all his stock. Goose bumps rose.

He talked about professionalism. He said we were part of the honorable profession of arms—a *profession!*—and we were to act accordingly. We were told not to lie, cheat or steal. Major Johnson harped on that. Honesty and integrity. He made a big deal about it. No one, he said, better be caught lying, cheating, stealing, or not acting with integrity. In fact, the word integrity was probably repeated at TBS more than any other word, with the possible exception of "Marine" (and "girls" and "food"). He said we were like the other professions, like law and medicine, which are specialized and have their own vernacular. He told us, therefore, to embrace, not cast aside, military-speak and all its creative acronyms and initialisms. He wanted us to speak in a language only Marines would understand, and to consequently appear indecipherable to outsiders. That was OK with him. That was truly being part of a profession, he thought. Then he wished us luck and ended the speech. The Student Company Commander rose and approached the Major. Silent words were exchanged. The Student Company Commander about-faced and yelled,

"STUDENT PLATOON COMMANDERS TAKE CHARGE OF YOUR PLATOONS AND CARRY OUT THE PLAN OF THE DAY!"

Which meant the next scheduled event, 10 minutes away. We were now officially "welcomed" to TBS. The first 20 minutes of a six-month-long schedule were done. As we were about to learn, each day of TBS was jam-packed with activity, forced activity, strenuous activity, cataloged down to the minute. I soon learned that the only way to survive, if not flourish, was to live in the present moment, living life minute by minute, if not second by second, not in the past or future. The schedule was too ominous and foreboding to do otherwise. *The Big Suck.*

PART 1: *THE BASICS*

This section describes the "basics" of The Basic School. In other words, it details, down to a granular level, the parts of TBS that are overlooked and boring but are very important to a lieutenant reporting aboard TBS. Some of these details are about the topography of TBS, including Quantico and Q-Town, checking-in, and wake-up times. Others are about procedures, like teaching methods, classroom instruction, and the roles of Staff Platoon Commanders. Still others are about the duties and responsibilities of TBS lieutenants, such as assuming student billets, fire watch, Officer-of-the Day (OOD) duty and paying for uniforms.

Checking-In

It was Sunday, September 22, 1991, around 1500 hrs, when I first drove onto Camp Barrett, where TBS is located at Quantico. The first thing that hit me was its diminutive size. Only a handful of buildings existed and the whole place seemed to sit in half a grid square (500 meters x 500 meters). I thought, *you got to be kidding me! This is it? Am I in the right place?* I was trying to find TBS Headquarters. My original orders, mailed to me in Michigan, stated that I had to check-in at TBS HQ from 0800-2400 hrs, a huge window of time. I didn't drive but a hundred yards, when I saw an uninspiring building with a motivating "Follow Me" statue sitting on a mound in front. It was TBS Headquarters, better known as Heywood Hall. I remember thinking, *the architect for the building should be shot, while the architect for the statue should be promoted and given a raise.* I pulled into a dirt parking lot—Lot 14—just across the road from Heywood Hall.

I was told many times that checking-in at TBS *would not* be like checking-in at OCS. It was supposed to be simple, easy, not traumatic. No screaming sergeants. No running around with full suitcases. Nobody in your face. Just a little nervousness…because…you never knew. After parking and waiting 10 minutes, contemplating my fate and what was to come, I got out of my Chevy Sprint and walked to Heywood Hall, orders in hand. I was casually dressed: khaki pants, short-sleeve shirt, loafers, short hair. As I opened the door, I was still thinking ALL HELL MIGHT BREAK LOOSE. I entered Heywood Hall on full alert; my eyes scanning left and right. No one was there. Silence. *Where was everyone?*

Checking-In

Was I the only one reporting-in? I was in a large lobby, with three hallways running from it. I picked one. Within 20 feet I heard, "Hey Devildog, over here." The sound came from a room I had just passed. I walked back to it and met a Captain. He was the OOD (Officer-of-the-Day). He greeted me respectfully, as if I was a full-fledged Marine officer even though I really wasn't. *Trainees Being Saluted.* He shook my hand and asked, "Got your orders?" I said, "Yes, Sir," and handed them to him. Quietly, he stamped, signed, and handed them back. He also assigned me a room in Graves Hall—a BOQ (Bachelor Officers' Quarters)—and said, "Take your orders and report to the Command Duty Office in Graves Hall, OK?" I said, "Yes, Sir." I returned to my car, drove 200 meters to Graves Hall, and reported to the Command Duty Office on the second floor. I found two student 2nd lieutenants standing duty. Come to find out, they were part of my TBS company. They were standing duty because they reported to TBS early (their OCS class had graduated a couple of weeks prior). They were, of course, friendly to me, because I was one of their classmates. After shaking hands and asking more than a few questions, like what was scheduled for the next day, and what was the Company Staff like, I was handed bed sheets, a pillow case, and a key to my room. They said my room was right down the hall. After I looked at the room, I went to my car and disgorged my personal effects and loaded them into my new home. But that was it for checking-in. Painless. Quick and easy.

During my TBS class, there were at least four companies in the training-cycle at TBS (approximately 220-250 Marines were in a company with five equally-divided platoons of 45-50 Marines in each company).[4] All four companies were at different stages. Some just reported aboard, and some were near graduation. Assignments to companies and platoons were neither random nor voluntary. I was assigned to Fox Company simply because it was the only "open" company ready to commence training. And I was assigned to 5th Platoon because of the alphabet. We were assigned a platoon in alphabetical order by last name, until a platoon had approximately 46 lieutenants; in my company, 1st Platoon contained lieutenants with last names beginning with A-Da, 2nd Platoon De-Ko, 3rd Platoon Hr-McV, and 5th Platoon Si-Z. Room assignments in the BOQs, as you would expect, were also alphabetical. Last names beginning with "A" had rooms at one end of the floor, while "Zs" were at the other. Everyone else was in between. In the beginning, three lieutenants were assigned to one room. As other companies graduated TBS, more rooms became available. We shifted down the floor into these new rooms, frequently going from three occupants to two.

---

4   The companies, named with military phonetics, ranged from Alpha, Bravo, Charlie, Delta, Echo, and Fox (short for "Foxtrot"). Another company, known as Mike Company, was used to house lieutenants who finished OCS and were awaiting the next TBS company to fill and start up.

Wake-Up Time

The average day at TBS began at 0700 hrs when we were in garrison and attending classes or going to the field for the day. Of course, that wasn't when we woke up; that was when our first class began. We arrived at class, depending how nervous the student billet holders were, anywhere from 0645-0655 hrs. During the beginning of TBS, much to my surprise, we couldn't walk to class individually like everyone does at college. We had to hold company formations with 230 Marines, with each of the five platoons reporting who was present or UA. For the first six weeks, we held these formations every morning, marching 150 meters to class in platoon formation, holding our books and notes, with the Student Platoon Sergeant issuing commands to the platoon. Holding company formations and marching platoons 150 meters takes more time than you'd think. So our formations began no later than 0640 on Chosin Ave, which was the road in between O'Bannon and Graves Hall, the BOQs. By the 7th week, because our company demonstrated promptness, we were allowed to forego the company formation and walk to class individually. But some lieutenants abused the privilege of walking to class individually by being late, so the company staff made us, once again, hold company formations and march to class. *The Big Shuffle.* This only lasted for a few days. After that, we were allowed to walk to class individually and never had to march to class again.

Now our wake-up time changed drastically if we were going to the field for two, three, four, or nine days. The night before going to the field, the company staff would post a list of lieutenants in each platoon-area of the BOQ. Lieutenants on the list were assigned a specific weapon to be drawn from the armory at 0430 the next morning. Yes, 0430. *Proud to serve!* Each lieutenant had to draw either a SAW (a 5.56mm caliber machine gun, known as a Squad Automatic Weapon), an M-203 40mm grenade launcher (which is a separate M-16A2, with the 40mm launcher—grenade tube—affixed under the fore-grip), an M-60E3 (7.62mm machine gun with a tripod and vertical fore-grip), SMAW (Shoulder-Mounted Assault Weapon, a "bazooka"), night-vision scope or binoculars. This unfortunately meant that most everyone had to be at the armory at 0430. Everyone then boarded a Cattle Car (yes, it looks just like a silver Cattle Car, fitting approximately 45 lieutenants most uncomfortably) at 0515 and were shipped to an immense training area, sometimes requiring a 30-minute ride. Which usually caused us to snore like thunder. Once in the field, our wake-up time didn't improve. Usually, we were awakened by the fire watch at 0400. Problem was, we usually didn't get rolling until 0600 or later. Why so early? Harassment Package, thank-you very much. *Time Being Squandered.*

Liberty

We had liberty on weekdays from the evening until the next morning (except for "remedials" that occurred after hours). We also had weekends off, starting from Friday evening and continuing until Monday morning (except for "remedials" that occurred on Saturdays, and one Saturday spent on the rifle range, and one Sunday spent on BASCOLEX). During the week, liberty commenced anywhere from 1500 to 2100. The exact time, however, depended on many things, like the time we returned from the field and finished cleaning weapons, or when our SPC decided he wanted to secure us. As can be expected, we were secured later during the first three months of TBS than the last three. During the first half, the average secured-time was 1930. In the last half, the secured-time was anywhere from 1500 to 1730. But each day was different. And sometimes, even toward the end of TBS, we got secured as late as 2000 hours. As always, the training schedule dictated.

TBS Topography: The Infrastructure

In 1991-92, only two BOQs existed at TBS: O'Bannon Hall, and Graves Hall, where I lived. Even though the layout of the rooms varied with each BOQ, they were essentially "dorm" rooms with big, ugly cinderblock walls. In Graves Hall, each room had a head (bathroom), one bunk-style rack, one pull-out couch, a window, two dressers, and two credenzas with pull-down desks. O'Bannon Hall was a little different. Each room in O'Bannon, while having the same furniture as Graves, was connected to another room through a head. Some lieutenants created a separate "sleeping" and "studying" area, by putting all the racks in one room and the desks in the other. But regardless of the BOQ, most of us rented small refrigerators, and had one phone per room (yes, a land-line phone, a now extinct dinosaur).

There was another interesting feature to Graves Hall, particularly to civilians who had never visited before. Each room had two individual rifle racks bolted to the concrete wall, just inside the door, holding one M-16A2 each. (If there was a third roommate, he chained his rifle to one of the two rifle racks). When civilians—read: girls—walked in, they looked down at the assault rifles bolted to the wall, and quickly realized they weren't in a college dorm, after all. They were in a manly village with machine guns strapped to the walls. Notwithstanding the curious looks from civilians, rifle racks in the rooms were a good deal for lieutenants. We used the M-16A2s so frequently at TBS, that it wasn't worth the time and effort to keep retrieving and returning them to the armory. Having them in our rooms, saved time—except at night. Every day, at the close of business, each platoon had to get an exact count of the rifles, by having one Marine physically ID and touch each rifle. They would only count the rifle if it was locked in the rack. But lieutenants were never back in the BOQ on time, certainly not all by the COB. Like anything else, some drifted in late, sometimes even legitimately. But no one got secured for liberty until an accurate count was made and "The Word" was given. The time it took to receive The Word always seemed interminable. *Try Being Secured.* The fire-team leader would check the rifles, then tell the squad leader. The squad leader would tell the platoon sergeant. The platoon sergeant would tell the platoon commander. The platoon commander would tell the company staff in a conference room down the floor…Then, after all that was accomplished, The Word would course its way back down, going through the same ears and mouths as it did on the way up. Each night. Without fail. Meanwhile, all the other lieutenants waited…and waited…and waited. You get the picture. Pain in the ass. *The Broken System.*

One last thing about the BOQs: the parking lots. Parking lots? Yes, the parking lots. No, they weren't filled with potholes, nor were they dirt lots (with the exception of Lot 14). As a matter of fact, the parking lots were nicely paved, nicely lined, and conveniently located right next to the BOQs. Problem? Someone never bothered to match the NUMBER OF TBS LIEUTENANTS with the NUMBER OF PARKING SPOTS. The numbers weren't even close. Way more lieutenants than spots. *This Base Sucks.* So when a lieutenant, at the close of business, after being secured, decided to drive off-base in his Lieutenant Mobile—as it was called—to have dinner, run a few errands, go on a date and the like, he returned to find no place to park. You could count on it. Where to park? Lot 14, which was 2-3 football fields away next to Heywood Hall, not the BOQs. God forbid the poor guy left and came back with bags of groceries. Or God forbid he left for the weekend and returned on Sunday night with a slew of things for the BOQ room. But even worse was the deterrent effect. These delightful little parking lots prevented us from making full use of liberty. You'd hear: "I don't want to go into town, I have a good parking spot." Parking lots—something so innocuous as parking lots—were part of the Harassment Package at TBS. They became part of the honor of saying, *"Proud to serve!"* a frequently repeated phrase by TBS lieutenants whenever they encountered something undesired or difficult.

Besides the BOQs, only a few other buildings existed at TBS. And some were used almost daily. One, as a matter of fact, was TBS Headquarters itself, Heywood Hall. We spent most of TBS in this building, because it housed four of the six indoor classrooms. The fifth classroom, Reasoner Hall, was the nicest of all classrooms, for it resembled a large movie theatre with nicely padded seats in rows sloping downward toward the stage in front. Also sloping downward were our eyelids every time we went there. Way too comfortable for TBS lieutenants. Maybe that's why we only went there two or three times. The sixth classroom was in the Infantry Officer's Course (IOC) building across from Heywood Hall and next to Lot 14.

There was also Ramer Hall—the natatorium—which contained a huge pool and gym. There was the armory, which housed one too many weapons which had to be cleaned one too many times. There was a small building called TSFO (Training Set Fire Observation), which was used only twice, once for simulated artillery call-for-fire, and once for calling-in simulated air strikes from simulated fighter planes. There were many Quonset huts nestled in and around the edge of TBS. Only a couple of them concerned us, however: the ones where we, during the first week, obtained all our gear, like flak jackets, ALICE packs (All-purpose Light Individual Carrying Equipment), E-tools (Entrenching tools), deuce-gear (cartridge belt with green suspenders), Mickey Mouse boots (air-filled Winter boots), wet weather gear, field jackets, galoshes, helmets, shelter halves, ponchos, sleeping bags, and so on. In addition to this gear, in a different Quonset hut, we picked up "Pubs," or publications, for the entire period of instruction at TBS. These were 8.5"x 11" paperback publications on every conceivable subject. Stacked on top of each other they easily measured four-feet (a lot of reading for a six-month course). We stacked them this way and inserted them vertically into our seabags and transported them back to the BOQs, usually in the back of pickups. During the last week of TBS—and what a glorious week it was!—we returned to give back what we were given. Except the pubs; they were ours to keep.

That was it for TBS, geographically speaking. We spent the majority of our time off TBS-proper and on immense training-areas, where we played war and immersed ourselves in nature and the elements—rain, snow, sleet, daytime, and nighttime.

Quantico and "Q-Town"

For those who want to become Marine officers, all roads lead to Marine Corps Base Quantico—known as "MCB Quantico"—which was formed in 1917 after the U.S. Government purchased the land. Quantico makes Marine officers and is known as the "Crossroads of the Marine Corps"—a phrase prominently displayed on the archway of the Main Gate. Think of the "crossroads" as the funnel through which all Marine officers must pass—not only to become commissioned officers but also to begin their careers and return many times afterward for more training and education.

Here's the thing about Quantico. There are two Quanticos: MCB Quantico, which is huge, and a town called Quantico, which is small and within the Marine base itself. The only access to the town, known as "Q-Town," is through one of three ways: the base itself, the Amtrak train station in Q-Town, or from the Potomac River. Q-Town, at first sight, seems like a Potemkin village. But upon closer inspection, one can see that it is, in fact, a real town with real restaurants, barber shops, laundries, and clothing stores; it even has a Post Office, not to mention the Amtrak station. Another important fixture contained within MCB Quantico is the FBI Academy. Many a Marine lieutenant has mistakenly wandered through the FBI Academy and its Potemkin village—Hogan's Alley—on land navigation exercises. When a lieutenant appears at the FBI Academy, in "boots and utes," it's pretty good evidence that he is beyond lost. That being said, at least he was still on MCB Quantico, and did not wander off-base in any of the surrounding towns of Garrisonville, Aquia, Dumfries, Alexandria, Washington, D.C., and Georgetown.

As for the educational part of MCB Quantico, it's densely packed with important and sometimes career-changing schools. At the time, in 1991, Quantico had eight Marine Corps' schools, and one school for civilians: the prestigious FBI Academy, as mentioned above. The Marine Corps' schools included two officer selection schools (Officer Candidate School and The Basic School); three Military Occupational Specialty (MOS) schools (Infantry Officers Course, Communication Officers School, and Computer Sciences School); and three career-level schools (the Marine Corps Staff Noncommissioned Officers Academy, Amphibious Warfare School, and the Command and Staff College). Because of these schools, Quantico witnesses the transformation of people who mystically and mythically morph from civilians to commissioned Marine officers and FBI Agents. Reduced to its minimum, Quantico contains the hopes and dreams of thousands of people who want to become Marine officers and FBI agents.

MCB Quantico sits in the northeastern part of Virginia. It's 35 miles south of Washington, D.C. and 70 miles north of Richmond, off the I-95 expressway. It is beautifully immense, covering 60,000 acres (roughly 90 square miles), and sits within three counties: Prince William, Stafford, and Fauquier. It has summers loaded with oppressive heat and humidity and winters with surprising occurrences of snow given the extreme heat and length of its summers. It has hill after hill of dazzling green foliage and deciduous trees. Matter of fact, there is green in every direction except down. When you look down, you see red. Red dirt. Miles and miles of it. If Quantico is short of anything, it isn't red dirt or green hills. Or nicely paved asphalt roads. I need to stop here and mention the roads. These serpentine roads, winding here and there, are lined with pine trees. But the roads do have one annoying characteristic. They have speed limit signs every 100 meters, or so it seems. One soon learns that a hallmark of Quantico—and all Marine bases, in fact—is the zealousness with which the Military Police enforce these speed limits. They are Road Nazis. Now, most Marines don't speed *intentionally;* they've learned not to, by word of mouth, personal experience, and because they're Marines who try to dutifully obey rules and regulations. But Marines do speed *unintentionally* (which isn't much fun, is it?) and even *intentionally* at times. The unintentional speeding is sometimes caused by "speed differential." Marine bases are famous for posting speed limit signs dropping the speed from 55 MPH to 25 MPH, with the 25 MPH sign often posted unseen around a curve, and with an MP sitting in a car next to it with a radar gun. This brilliant placement of signs and MPs usually causes a driver to lock the wheels, skid, swerve, nearly run off the road, and have an out-of-body experience—all in the name of obeying the speed limit. Now at other times, the 25 MPH sign is posted on a straight road, in plain view, for at least 100 meters. While this placement is usually much safer, it presents another conundrum for Marines. It causes them to wonder, *Where* does the 25 MPH speed limit take effect? Right at the sign? Some distance *before* the sign? When you first *see* the sign? Only *after* the sign? Marines actually talked about this. All this reveals one important fact, which will become clearer by the end of the book: Only Marines can be this anal retentive.

There were two exits off I-95 for Quantico. The first exit, if one is traveling south from Washington, D.C., is titled, MARINE CORPS BASE-QUANTICO. Marines called this the "Front Gate." (The word "gate" is a military term of art, but there is no such thing. There is no gate, no guard shack, no sentry.) At the end of this exit ramp, there is a military-looking—could you have guessed?—green sign, with approximately 20 horizontal line-items, and an arrow next to each one telling you—*ordering you!*—which way to go. OCS to the left, near the Potomac River, and TBS to the right. (OCS and TBS are separated by a 20-minute car ride.) The second exit is titled, GARRISONVILLE, and is four miles south of the first exit. It leads you to the—what else?—the Back Gate, which is closest to TBS. This second exit, of course, contained no actual gate, no guard shack, no sentry, and not even a military-looking sign. At the exit, you turned right, traveled through burgeoning Garrisonville, turned right on Onville Road, and proceeded three miles until TBS (Camp Barrett) popped-up on the right.

Quantico also is situated in the middle of history. Not only is it located between the two Civil War capitals, it's surrounded by some of the most famous battlefields of the Civil War: First Bull Run, Second Bull Run, Fredericksburg, Chancellorsville, Spotsylvania, The Wilderness, The Valley Campaign, Petersburg. These fields of blood are tucked between the two capitals, which are only 105 miles apart, and are all within a short driving distance of Quantico. An hour-and-a-half drive separates the North and South capitals. Simply amazing. Also amazing is that the Commonwealth of Virginia, known as the "seat of the confederacy," and thought of as a wholly "southern" state, is contiguous to the Union capital. It runs right up to Washington, D.C., and even encompasses a portion of the modern-day Beltway. In fact, there are portions of Maryland, a true northern state, which extend more to the south than portions of Virginia. If you look east from the upper third of Virginia, you'll see Maryland across the Potomac—so much for being a "southern" state geographically. But make no mistake, Virginia *is* a southern state (or rather commonwealth) to its core; just consider the accents and dialects, the manners, the courtesy, the polite drivers who don't honk even when they're about to collide with you, the omnipresence of the Stars and Bars, and the military tradition, which may be its *raison d'etre*.

SPCs and Not-So Bright

At TBS, platoons were not commanded by enlisted personnel. They were commanded by Marine officers, otherwise known as Staff Platoon Commanders, who were either captains or 1st lieutenants. They were the "best of the best" in the Marine Corps. And they were known by the acronym, SPC, except if you addressed them in person. Then it was "Sir." The theory for having officers as SPCs was to provide experienced, top-flight officers for TBS lieutenants to observe and emulate on a daily basis. Unlike OCS, where we were "candidates," we were now lieutenants, i.e., officers. A good way to learn how officers lead is by observing and mimicking them. Observing our SPC's attributes and characteristics helped us to form our leadership style—what John Keegan appropriately called, and what he titled one of his books, *The Mask of Command.* As neophytes and acolytes, we honestly didn't know what constituted good, competent leadership. That's why observing and emulating an SPC, at least his good attributes, was so essential for us. We then shamelessly adopted the leadership style. As an extra bonus, TBS provided to my platoon, a legend. My SPC gave proof to the notion that superb leaders don't always fit the stereotypes; they come in all shapes and sizes.

His name was 1stLt James M. Bright. And if anyone was more idolized and revered than our Company Commander, Major Johnson, it was him. He was from Meunster, Texas, and was equipped with a moderate Texas drawl and a basic vocabulary. He was 27, 6-feet tall, wore nerdy spectacles, wrinkled cammies, and a Bronze Star with combat "V" (for valor) for his exploits in the 1991 Persian Gulf War. He was awarded this medal of valor in a ceremony in front of our company. Colonel Fawcett, TBS's Commanding Officer, read the battle citation which said that Bright exposed himself to murderous Iraqi machine-gun fire and saved a couple of wounded Marines from his platoon. For the remainder of TBS we noticed that other SPCs and Instructors would walk by him and affectionately call him Hero, and he would snicker, look down, and return with a smart comment. They would also call him "Not-So." A not-so reverent sobriquet for "Not So Bright." Brave but not so bright. (Don't think *we* ever called him that; oh no, that wouldn't have been good.) Sometimes, when we were with him in the field, the field radio—a PRC-77, known as a "Prick 77"—would hiss and crackle. We'd hear, "NOT-SO!…ARE YOU THERE?…OVER." Whereupon Bright would pick-up the handset and answer, just as if called by name. Let me tell you, by the end of TBS, my platoon deified him. He was our own legend. Matter of fact, he had more effect on me than any other single person at TBS. He broke the mold on Marine officers and demonstrated that good leadership comes in all shapes and sizes, wrinkled cammies and starched cammies, nerdy spectacles and perfect vision, a small vocabulary and large vocabulary, an extrovert or an introvert.

We first met Not-So at the close of business on our first day. Our billet-holders told us to assemble in the conference room (lounge) of our BOQ. Four or five couches were against the perimeter of the room. We were sitting on the perimeter, either on couches or on the floor, backs against the walls. We waited. No Not-So. Then, after a few minutes, he opened the door. We bolted to our feet, and someone screamed, "ATTENTION ON DECK!" Not-So walked in and said, "At ease, Devildogs. Take your seats." He walked to a couch. "Get off my couch," he said, and the couch emptied. Not-So sat down, looked at some papers, and then looked up and around the perimeter. We stared back, all 46 of us. He was, like us, wearing cammies, but his were noticeably wrinkled with up-turned collar-ends, which hid his lieutenant's bars. Collars flapping in the wind. He introduced himself and said he was our God for the next six months, and so on. I was listening for nervousness, stress in the voice, unsteady eye contact. None. It was immediately obvious that Not-So was an experienced leader who was comfortable with command. His command presence was phenomenal. He had a strong, calm manly speaking voice, with a Meunster twang. He was not exceedingly articulate like Major Johnson, but he was a good speaker who held your attention. You could tell he focused on details. He talked about the details down to a subatomic level. He made sense. He gave real-world examples, battle examples. I could tell straight away that Not-So was pretty bright, which was consistent with his surname but contrary to his nickname. He had been in the Corps for four years, and had been all over the globe, commanding platoon after platoon. Heck, he was a war hero. That was enough for us.

Not-So didn't speak too long. He introduced himself, and then he laid the groundwork, the rules, what the Marine Corps expects, what he expects. One thing I remember well. He asked us, "Why stand when you can kneel? Why kneel when you can sit? Why sit when you can lie down?" His point was this: go all the way. No half-assed work. When you were told to work, you expend 100% effort, nothing less. When you weren't supposed to work, you expend 100% effort in relaxing if not sleeping, regardless of where you were. *The Body Softener.* He told us not to "pretend" to look busy. He said we wouldn't fool him, nor would we fool anyone else, in the Marine Corps or even in the other world, which felt like galaxies away, the civilian world. He said there would be much down-time in the TBS schedule. So you must conserve your energy, he said. There is more than enough up-time. After enlightening us with that, he ended. He stood up. Someone yelled, "ATTENTION ON DECK!" And Not-So said, "Carry on Devildogs." He walked out and we returned to our rooms.

At the end of the first week, on late Friday afternoon, Not-So told his student billet-holders to assemble the platoon outside. It was time for a Liberty Brief; we were on the cusp of being released for our first weekend of liberty. We assembled outside the BOQ, next to the parking lot. We waited in platoon formation for Not-So. After "smoking and joking" for a few minutes, he appeared. The student platoon commander called the platoon to attention, and Not-So said, "Stand easy, Devildogs." He stood, centered and alone, about 10 feet in front of us—still wearing wrinkled cammies, with curled collars, and his hands deep in his trouser pockets, which is a Marine no-no. Marines weren't supposed to put their hands in their pockets. They even derisively called it, "Army gloves." Not-So couldn't be bothered with things like that. He let loose the Meunster twang.

"All right people, listen up...Real quick... This is a Liberty Brief, the first of many, and it's going to be brief. Over the weekend, don't drink and drive. Don't drive and drink. And don't suck nobody's dick.

Student Platoon Commander, take charge of your platoon."

By the time he said, "Take charge of your platoon," we were hooting and hollering and ooh-rahhhing and generally losing our marbles. We looked at each other, thinking *Did he just say what I think he said?* The man, the wrinkled war hero, was just a scream—complete with double negatives. We wanted to bronze the guy. His Liberty Brief made the rounds and circulated through the company in no time. Not-So was adding to his own legend.

Teaching Methods

After the first couple of months, I concluded that TBS was a high-quality school with high-quality instructors. To be sure, the instructors—usually captains and majors—were at times autocratic, uncaring, and loud. *"Lieutenant, you see what you did with that slick maneuver?...You killed all your men!...Every last one of them!...Beautiful!...Are you an idiot or what?" Totally Biting Sarcasm.* But for the most part, they were passionate, competent, remarkably effective teachers, who wanted us to learn and comprehend the subjects being taught, which usually began and ended with tactics: squad, platoon, offensive, defensive, and weapons. *Tactical Boy Scouts.*

At TBS we were taught these tactics—as well as most everything else—through a combination of methods: classroom instruction, sand-table exercises, and practical application ("Prac Ap") exercises in the harsh environment known as "the field." The teaching and learning of tactics progressed in a logical way at TBS. We first participated in a STEX, then a TEWT, and finally a FEX. A STEX? TEWT? FEX? *Tewts, Boxes, Stexes.*

A STEX is a Sand Table EXercise. STEXs were taught to us in Classroom 4. In the middle of CR4 stood a huge rectangular sand table, surrounded on three sides by aluminum bleachers. The tactics were first introduced to the company on this big sand-table. Then we broke into groups of 8-12. These small groups went to the perimeter of the classroom, where approximately 20 smaller sand tables stood along the walls. A captain or a major led these more manageable groups in a free-wheeling tactical discussion, a tactical scenario. The instructor wanted—demanded!—participation from all the students. In fact, he usually picked three of us to assume the role of company commander, platoon commander, and squad leader. Then we got to play in the sand. We moved plastic men, material, and weapons to various spots, until we were satisfied. We placed these figures in positions we thought advantageous, given the tactical scenario before us. We had to explain *why*. Then we got critiqued. After the critique, we moved outside to Training-Area 8 (TA-8), the area surrounding TBS-proper, to participate in a TEWT.

A TEWT is really a misnomer. It stands for Tactical Exercise Without Troops. But you see, there *were* troops. *We* were the troops. And, believe me, we were there and participating in the tactical exercise. A TEWT seeks to accomplish putting newly taught tactical concepts to use by—and this phrase gets worn out by the end of TBS—"walking the dirt" or "walking the dog." One must "walk the dirt"…"walk the dog"…to get a feel for the lay of the land, to see the undulations, the denseness of the forest, so one could effectively employ the equipment and Marines provided him. The TEWT, as with the STEX, was usually taught by a captain or major. He explained the situation to the group—usually platoon-sized, but sometimes squad-sized—then allowed us to wander around the dirt, deciding where we would specifically employ our SAWs, M-60E3s, M-203s, and fire-teams. We usually were assigned the role of squad-leader, but at times we were the platoon commander. After wandering and deciding on specific locations for men and weapons, we returned to the instructor (all officer instructors were called AIs, for Assistant Instructors; all enlisted instructors were EIs), who would then pick one of us to present a detailed plan to the platoon. Spears and arrows were launched. *Tactics Being Slaughtered.* All the other lieutenants, not to mention the AI, got to critique the poor guy's plan. Most of the AIs were courteous and wouldn't excoriate him for having a bad idea. To be sure, there were some instructors who would take egotistical pleasure in demeaning a new lieutenant. But the clear majority honestly believed there were no "right answers"—as long the plan was backed with a reasoned and logical analysis.

A FEX is a Field EXercise. In the learning progression, it followed both the STEX and TEWT. In short, it was a deployment to the field to employ tactical concepts, such as offensive and defensive tactics, patrolling, convoy operations, attacking a fortified position, and so on. Some FEXs lasted only a day; others lasted two, three, and four days, with the 9-Day War as the epoch. If we went on a one-day FEX, we returned to the BOQ anywhere from 1500-1800 hrs. If we went on a two-, three-, or four-day FEX, we stayed overnight in the field. Our days in the field usually ended around 1700 hrs, whereupon tents were pitched, fires started, and stories of the day recounted, modified, and glorified.

For most FEXs, we had to draw weapons—SAWs, M-203s, M-60E3s, SMAWs—from the armory. We also had to draw binoculars, night-vision goggles (and scopes), and radios. But the drawing part wasn't so bad. It was the returning part which tested our patience. The weapons, after dragging them through the red mud of Virginia and firing countless blank rounds through them, were filthy. Filthy beyond description. It also didn't help to have the BFAs (Blank Firing Adaptors) screwed into the muzzles of our weapons. BFAs only added to our misery. They stopped most of the gases from the blank rounds from leaving the muzzle, which turned the insides of the weapons a nice shade of black. BFAs acted as a porous cork at the end of the barrel and were a necessity for gas-operated semi-automatic or automatic weapons firing blank rounds. BFAs did more than just stop the gases at the muzzle; they routed the gases back through the weapon's gas tube, which ultimately pushed the bolt back, extracted the spent round and chambered a new one. Without a real projectile sealing the barrel on its way out the muzzle, the only way to send gases back to the bolt was through the use of a BFA. So as you can imagine, thanks to the red mud, blank rounds, and BFAs, our weapons were black. Blacker than black. Cleaning them, therefore, was never quick and easy. And trust me, we weren't going anywhere until that armorer—usually a lance corporal—put his tiny initials on a card, signifying that the weapon was clean and OK to be returned to the armory. We hated this. Frequently our entire company would return in the late evening from a two-, three-, or even four-day FEX on a Thursday or Friday night. But before any lieutenant could get secured for the night, all the weapons checked-out from the armory had to be returned. Cleaned. And spotless. *Try Being Secured.* So when we made plans before the FEX, we were wary. It always required an extra three hours at the armory once we returned from the field. And we never knew the exact time when we would return from the field. Because of this, many lieutenants would "conserve" their ammo by not firing their weapons; they didn't want that powder piling up, causing them to scrub, scrub, scrub. They wanted tiny little initials on their cards as quickly as possible. Once, one of the guys in my platoon, 2ndLt Jim Titcombe, was supposed to pick up his girlfriend from Washington National Airport—on a Friday night, after having been in the field for three days. Do you think that lance corporal at the armory cared? Didn't matter to him. Titcombe arrived at Washington National over two hours late. *Temporarily Behind Schedule.* He was greeted with bags—thrown right at his chest.

The majority of FEXs were platoon tactics—both offensive and defensive. One of the first FEXs was "Squad-a-Thon," a two-day evolution employing a squad in the offensive. For this FEX, we were picked randomly to play the role of squad leaders, to receive and issue a 5-paragraph order, and to lead our 13-man squad to the objective. This exercise was similar to SULE II (Small Unit Leadership Exercise II) at OCS, except it wasn't graded and weighted—neither were any of the other FEXs, TEWTs, and STEXs—and there was no tortuous 15-mile hike to get to the training area. Almost every one of us was picked to lead the squad over the two days. Only one or two of 13 lieutenants were successful in dodging the icy glare and pointing finger of the AIs (the TEWTs were a little different: only infrequently did you get "volunteered"; mostly it was the *other* guy who got "volunteered"). All the FEXs worked this way. Except that FEXs involved leading more than just squads. They usually required leading platoons of 45 men or patrols of 15.

Teaching Methods

Did I mention terrain models? Before every patrol and platoon FEX, when we first arrived at the training-area, we had to build a terrain model—whether we liked it or not. The student squad-leader (or platoon commander or patrol leader) picked another lieutenant to help him. A terrain model is an area on the ground that is boxed-out (approximately 8'x10') and accurately reflects, on a much smaller scale, the terrain the unit would be passing through to get to the objective. Hills were shown by piling dirt or leaves or snow. Rivers were shown by using blue yarn or blue spray paint. Units and "attack points" and "assembly areas" and "lines of departure" and "objectives" were shown by writing on 3-by-5 cards. Draws and valleys were shown by digging out dirt. When the unlucky designee studied his map long enough and put the notable features inside the terrain model, he issued the 5-paragraph order (or patrol order) to his men while standing over this model. He walked through it, pointing and touching different areas for emphasis, as he was speaking.

You might ask: How did we know which side won the FEX? Or more specifically: How did we know when the enemy was "killed," or when *we* were killed? Simple answer. MILES gear. MILES gear was the Marine Corps' equivalent of Laser Tag. Except that MILES gear was outdated, heavy, burdensome, didn't work real well, and was hated by all lieutenants. It was much too cumbersome and heavy. Do infantryman want to add to their combat load? Isn't a helmet, flak jacket, deuce-gear, combat boots, M-16A2, and ammunition enough? Of course, they are. But MILES gear was yet another part of the Harassment Package. *Tortured By Sadists.*

MILES (Multiple Integrated Laser Engagement System) gear comprised three pieces of equipment. One was the transmitter itself, a rectangular box fastened to the top of the muzzle of the M-16A2s. The second piece of equipment was the "helmet band" wrapped around the outside of our helmet. It had maybe five protruding receptors—nodules— on the band, with a metal rectangular box on the back. Its drawback was also its weight; especially after hours of wearing it. There was a big difference on our neck muscles between an unencumbered helmet and one fitted with the MILES box and band. The third piece of equipment was a funny contraption. It mostly resembled a vest made of vertical and horizontal suspenders worn over our flak jackets. The vertical suspenders contained numerous receptors, and on the back, along a horizontal suspender, there was yet another large pain in the ass: a rectangular metal box—roughly 5-by-7-by-2 inches. Although it was heavy, its main drawback was its size (ever try to lean your back against a tree, or lie on the ground, with a big obstructionist shoe box on your back?) What did the box do? It hummed. It was all for humming. After you were shot by an M-16A2 transmitter, and your receptor was hit, you hummed. You turned into a loud nuisance, which sounded something like an otherworldly intergalactic hum. It was loud. Hmmmmmmmmmmmmmmm! And you could not turn it off. Only an EI with a magical key could do that—and only when he wanted to. EIs usually came around at the end of the battle. Meanwhile, until they did, Marines just got comfortable. And I mean comfortable. During FEXes you could look across the fake battlefield and not really see anything. You'd just hear Hmmmmmmmmmmmmmmmmmm! Everywhere. Like it was closing in on you. Hmmmmmmmmmmmmmmmmm! Hmmmmmmmmm! Then you would squint, focus some more, and see camouflaged anthropomorphic lumps here… there…everywhere…spread all over the ground humming away. If you looked closely, you saw these lumps lying on their backs, hands interlocked behind their heads, twigs in their mouths, resting, loving life (the lumps thought getting shot was much easier than attacking a fortified position). Christ, some lumps even slept right through the damn Hmmmmmmmmmmmmmmmm! which was wailing from the pain-in-the-ass-box right behind them. Zzzzzzz zzzzzzzzzzzzzzzzzzzzzzzzzzzzzzz! (TBS lieutenants can sleep through anything.) But most of the lumps just waited, and grabbed chow from an MRE, rolled from their back to their side, from their side to their stomach, and from their stomach to their back again…Humming away… Hmmmmmmmmmmmmmmmmmmmmmmmmmmmm!… eating, picking their teeth…Not one care in the world, unless the lump rolled over into another lump…of human excrement. Which is not uncommon at Quantico. After all, do you think Porta-Potties are available in the field? In any event, after 10, 15, 20, 30 minutes, you'd see an EI walking from lump to lump, bending over, inserting a key into a mechanism on a MILES vest, twisting, turning off the miserable little Hmmmmmmmmmmmmmmmmmmmmms. Gradually, one by one, the hums died away. Hmmmmmmmmmmm! to Hmmmmmmmm! to Hmmmm! To Hmmm! To Hmm!…To peace and quiet…To a body-count check…To another FEX.

The best part of FEXs were the evenings. After hiking, defending, attacking, patrolling, and digging fighting holes, an approaching sunset was always looked forward to. After a long day, we picked a clearing for our company and lined up the platoons and tents. We created working parties to collect dry wood and kindling for the campfires. Campfires were an endless source of pleasure and entertainment. We would sit around them, dry ourselves, cook our meals, and recount funny stories of the day. We would sit and stare at the flames, watching the molecules turn into gas and heat, mesmerized by the yellow and red and blue flames, the twisting smoke, which sometimes blew right into our nostrils causing us to look away in pain and to cough…the crackling *pop*…*pop*…*pop*, *pop* of the wood. And then the stories themselves. You couldn't help but have fun. Everyone was happy just to be done for the day.

Campfires were especially great in the winter. Northern Virginia got pretty cold, sometimes with five or more inches of snow. In this kind of weather, campfires were invaluable. I remember one FEX in the middle of winter. It was maybe 20 degrees. An inch or two of snow was on the ground. After tromping through the woods all day, everyone's feet were cold and numb. We sat around the campfire. Some Marines took their boots and socks off and placed them by the fire. Most others, including me, didn't remove their boots, but rather clasped their hands around their knees and shins, placing the bottoms of their boots at 45-degree angles near the fire to dry them first. On this day, I had earlier forded a stream. Unlike regular combat boots, my boots, which were called "Rocky Combat Boots," were insulated and would sop up all the water. They were terrific boots, but only if you kept them dry. They cost $120.00 back then, which was a lot. While sitting there, just trying to relax before I took my boots off to dry my socks and feet, I heard some Marine yell, "DUDE, LOOK! HIS BOOTS ARE MELTING!" Immediately, I looked around to see what idiot put his boots too close to the fire. But all fingers were pointing to my boots. And so were all eyes and laughing faces. I then heard, "MAN, THEY'RE MELTING! Heh heh heh heh heh heh!" I thought they were joking. I mean, my feet weren't even warm yet. But then I looked at my boots. They were smoking. That was enough evidence for me. They obviously weren't joking. It was time for quick thinking. It was time for quick action. In my view, the best way to put out a burning boot was to smother it. So I stood and clomped around on the cold ground. In hindsight, I should have sat there. At the time I stood up, the bottoms of my boots were molten rubber, like a dragster tire, almost no waffle pattern in sight. After walking around for less than a minute, I returned to my original position. Twigs and grass were now cemented into my soles as if planted there, sticking out at odd angles and lengths. I looked down. "Oh, my!" was all I could muster. Uncontrolled laughter was unrolling around the campfire, which was quickly spreading to other campfires. I sat down and tried to pull out the twigs and grass, but they wouldn't come out. The rubber had already dried from walking around on the cold ground. Marine improvisation was next. I pulled out my KABAR and executed a militarily precise trim job around the edges of my soles. I made those $120.00 Rocky Combat Boots almost new again—except for the flat bottoms with embedded twigs and grass.

Classrooms

Many Marines, usually the infantry-types, hated classroom instruction because they wanted to be in the field, walking the dirt, playing war. But there were good things, entertaining things, about attending class. Each of the four main classrooms in Heywood Hall had a large movie screen on the forward wall. Once we were seated for instruction, and sometimes even as we entered the classroom, a violent movie clip would be playing, such as real combat footage from an air raid during a war. No sound would come from the movie. The only thing you heard, and believe me it was loud, was blaring rock & roll music, like Van Halen, AC/DC, and Guns & Roses. Commingled with the music were lieutenants ooohh-raahhing at the sight of bombs bursting, people dying, buildings disintegrating. The macabre Marine humor. The video and music were called Attention Getters, and they worked. It was, by far, the most enjoyable part of classes. Some instructors even took it a step further. After the short video, some would grab a book on Congressional Medal of Honor recipients and read battle accounts where the CMH was awarded. At the end of the recital, lieutenants oohh-raahhed without interruption. And then the class began.

Classroom instruction was fairly typical. Instructors usually conveyed information—known as knowledge for the "knowledge knot" or "brain housing group"—through lectures and slide presentations. Because the classrooms were huge, typically containing 250 lieutenants, each instructor usually wore a tiny microphone and carried a remote control for the slides. The instructor would walk up and down the center aisle, talking and clicking the remote…talking and clicking, hour after hour, slide after slide. We grew weary of slide presentations after a couple of months. *Thousands of Boring Slides… Thousands of Blank Stares.* And don't think these blank stares weren't noticed by the instructors. The instructors saw the tired heads falling forward, like chickens pecking corn, the heads jerking back, swiveling around to see if anyone noticed; they saw the drooling. Oh sure, they saw all this. They knew. They were graduates of TBS themselves. The only issue was how to deal with it. Some instructors— but only a few—would simply ignore the pecking, the nodding, the sleeping, the blank stares, the drooling. But most weren't that nice. Most would seize the opportunity and cajole, embarrass, bother, mortify, and excoriate the poor soul who dozed off and got caught. Marine sadism at its best. *Tortured By Sadists.* Here's an example of what I mean, and it unfortunately occurred more frequently than we had liked. When a lieutenant dozed off, the instructor walked down the row, crouched in front of the sleeping lieutenant, and waited. And waited some more. Everyone's heads swiveled around and focused on the poor guy, like 229 spokes leading to a hub. The whole subject matter of the class became the duel. Within a minute, the class began to laugh, which awakened the poor soul. When he opened his eyes, he saw a nice Instructor Face not one foot away. Smiling. You could feel the poor guy's heart pounding. His eyes got bigger than big. He sat up. His mouth went oval. One hand went up to wipe drool. The Instructor Face took over from there and proceeded to publicly embarrass him, asking him why he wasn't paying attention, telling

him how disrespectful he was, sleeping…drooling even…
while someone was lecturing. *Totally Biting Sarcasm.* Finally,
much to the poor guy's relief, he was ordered to the back
of the room—to stand for the remainder of class—while
the Instructor Face continued with the lecture like nothing
had ever happened. Now, of course, that was just one way
of handling it. Other instructors took other measures, like
a baseball bat, or some other intimidating implement, and
swung it like an axe, exploding on the desk inches from the
sleeper. The instructor would grin and point to the rear of
the classroom. Off went the sleeper. Some instructors were
a little nicer and would walk over, stand by the sleeper, and
whisper in his ear. Sweet little nothings. Off to the rear
he went. But the most common form of confrontation
was this: An instructor, while walking and talking, would
suddenly, out of nowhere, point a perfectly straight index
finger at a lieutenant who was dozing, from far away, and
say, "Lieutenant, grab your trash and get to the back."
After witnessing a few of these encounters, we'd elbow our
brethren who were dozing off. But after a while, most didn't
need the elbow. They'd realize their eyelids were drooping
and would stand up and walk to the rear of the classroom.
Marine initiative. Marine discipline. Stand-up guys.

Attending class, as you can imagine, made us feel
edgy. Who wants to fall asleep and be embarrassed in front
of 229 peers? Another thing that made us edgy was getting
called on. It was similar to law school. Most instructors
made the Socratic method a major part of their instruction.
When an instructor stopped talking, and started scanning
the room, looking down each row for a victim—well, you
knew trouble was brewing. You tried to put cover and

concealment to good use: you looked down, avoided eye contact, tried to look busy, took notes, the whole bit. But the index finger, like a pistol, would hit one of us. After being shot, we stood and said, "Sir, Lieutenant Smith, 5th Platoon. The answer is…" Now, if the answer was correct, no problem. We just took our seat again, eased back into quiet concealment and oblivion, knowing we wouldn't be called on anymore—at least for that class. But if the answer was wrong—oh boy. Here we go again. More sarcasm, more laughing. The macabre Marine humor erupts. But it wasn't just the answers you had to worry about. The sarcasm and laughing also applied to asking questions. Stupid questions. They too received catcalls from the class, and opprobrium from the Instructors. *Tension Building School.* But again, this was not universal. Some instructors were merciful and wouldn't deride us for asking a dumb question or giving a dumb answer. It all depended on the Instructors, just like the TBS experience itself. Your experience depended on your Instructors and Staff.

During most classes, we would hoot, holler, and howl at things said by instructors. And the subjects we howled at were revealing. For example, we'd hoot, holler, and howl at anything about physical pain, toughness, adversity, killing, maiming, or the Marine Corps itself. It was part of the masochistic and macho culture. Love pain. Love anything that hurts us or the enemy. *The Bleeding Sphincter.* We knew our opponent would not love pain—at least not like we loved pain. Pain and adversity combined to form the sea Marines swam in. We'd ooh-raahh when an instructor told us about a grueling PT event coming up. Or about anything that would cause lieutenants' pain.

We took more than just knowledge from the classrooms. We took snappy words and phrases from the Attention Getters and Instructors. For weeks on end, for no apparent reason, we'd repeat these words and phrases. In the field, in the BOQ, in the classrooms. It didn't matter. During one class, the Attention Getter involved a clip from the movie *The Untouchables*, where Al Capone, played by Robert DeNiro, was having a meeting with his *capos*. DeNiro got up and walked around the table. He picked up a baseball bat and talked about the importance of operating like a team, having all the players contribute, and the like. He was gesturing with the bat, saying "We got to be a team…T-e-e-e-e-e-eeam!… you understand?…T-e-e-e-e-e-eeam!" And then he gave his best homerun swing, in a nice game of T-Ball, smashing one guy's head like a cantaloupe. Our classroom went nuts seeing the guy's head explode! Barking! Ooh-rahhing! 150 decibels! Marine humor! After the class, again for no apparent reason, we chanted "Teeeeeeeam!" in the hallways, outside, in the BOQ, everywhere. During other classes, out of nowhere, we would chant "Teeeeeeeam!" interrupting the lecture, with the Instructor wondering what the hell was going on.

Another example involved a class on terrorism. An instructor told us how important it was to vary our routines. It helped prevent terrorist attacks and getting killed. He gave an example of someone who didn't vary his routine: Aldo Moro. Aldo Moro was killed by terrorists in Italy because his routine was too routine. Again, I can't explain why, but lieutenants ran with the name, Aldo Moro. After the class, you'd hear it yelled, but not fully. One lieutenant would yell, "ALDO!" and another lieutenant would answer, "MORO!" You'd hear it in the classroom, you'd hear it outside, in the parking lot, in the field. For example, we'd be standing around a campfire and hear, "ALLLLDO!" out of nowhere, from a source hundreds of meters away. Then of course, also out of nowhere, someone would answer, "MOOOOORO!" And they'd repeat it, like a musical refrain, over and over, until they got tired, and until everyone else joined in. We'd just laugh. This happened so many times, with so many different words and phrases, that it's hard to remember them all. It was part of the culture, part of relieving stress, part of having fun, part of the macabre Marine humor.

The last thing I remember from classes was the odor. Now. This odor didn't emanate from the building or the walls or the desks. It emanated from the lieutenants themselves. No, it wasn't B.O., for lieutenants usually kept themselves clean and smelled like cologne or soap, except in the field. The smell was from farting. And let me tell you, there were times the farts were so bad they burned our noses. Truth be told, Marine lieutenants are world-class farters (and burpers, too). And they have no stage-fright. They can unleash their chemical warfare anywhere. Doesn't matter. In class, in the BOQs—*These Barracks Smell*—in the field, during a forced march, where it can be so bad, even outside, that you feel like breaking off from a column of twos and running into the woods to get fresh air. Interestingly, and this applied more at OCS than TBS, the farts could smell the same. After all, what's the computer initialism GIGO, Garbage In, Garbage Out? Well, when you're human and you eat the same food, or nearly the same food, your farts smell the same. I remember thinking—*wow!*—that smells just like my fart. GIGO. It almost made you feel like you had a proprietary interest in your fart.

Billets

The Marine Corps is big on management and leadership—*managing* material resources and *leading* men and women. Even though human assets are technically resources, and are subject to management principles, leading is what the Marine Corps wants you to do with men and women. Think of the difference between management and leadership this way: One doesn't *manage* his men up a hill to take a machine gun nest and accomplish the mission. One *leads* his men up the hill, from the front, takes the hill, then accomplishes the mission. Let's call the duties and responsibilities inherent in management and leadership "the art of command."

At TBS we learned the art of command principally through the assumption of "billets." Billets were positions of command, assumed for a set period of time, where lieutenants got to boss around other lieutenants. These billets corresponded to military rank. In the Fleet Marine Force (FMF), one particular rank was usually associated with one particular billet. So at TBS, we wore the rank associated with the billet—even if it was an enlisted rank. We wore the rank device on a square piece of red or yellow felt (red for platoon; yellow for company), and pinned it to our collars, exactly where our lieutenant's bars were supposed to be. The red and yellow felt made billet-holders stick out. And like the old Chinese proverb: the nail sticking up the highest gets hammered the hardest.

There were two groups of student billets: company and platoon. Both types were two-week assignments. The

Billets

company billets included Company Commander, Messenger (the Company Commander's "water boy," so to speak, who after two weeks, became the Company Commander, so getting this billet was really a four-week billet), Executive Officer, Company Supply Sergeant, and Company First Sergeant. The platoon billets were Platoon Commander, Platoon Sergeant, Platoon Guide, Squad Leader, and Fire-Team Leader.

Billets, putting it simply, were not desired—especially the company billets, for they required commanding up to 230 lieutenants, as opposed to 45 in a platoon billet. They lasted a considerable period of time, they were stressful, they required much extra work, and they never seemed to help you get through TBS. They only plagued you—and potentially dinged you. *The Big Suffering.* The assumption of billets was a duty. But thankfully, the duty didn't occur too frequently. During all of TBS, I had no company billets, one Platoon Sergeant billet, one Squad Leader billet, and maybe eight Fire-Team Leader billets.

Fire Watch and Fire Alarms

Fire Watch? To civilians, the words sound funny. In the military it's anything but funny. "Fire watch" was a euphemism for "Harassment Package." It meant guard-duty while everyone else slept. I guess the term was created in the old days when soldiers were instructed to watch the campfire to make sure it didn't die-out or spread-out. In modern days, at OCS for instance, it became part of the illustrious Harassment Package. Two candidates, on two hour shifts throughout the night, armed with flashlights, dressed in cammies, combat boots, covers, and web-belts, walking around the barracks, with no campfire within miles, looking exceedingly stupid and feeling totally useless. It was that simple, and that purposeful. At TBS, thankfully, we only had fire watch once or twice—and only in the field, where there were campfires. It was probably because we were Marine officers now, and not candidates. But don't think we got away without being harassed. TBS had something instead of fire watch: anonymous fire-alarm pullers, who would love to yank them in the middle of the night. Once the alarm was activated, we had to muster outside the BOQ and wait 10-15 minutes for it to shut-off. In the middle of the winter. At 0230 hrs. It took one or two of these delightful sessions to finally figure out that Graves Hall was not burning down. So we had an internal integrity battle: did we acknowledge the alarm and leave the BOQ as we were previously told to do, or did we roll-over and wait for our other alarm, the alarm clock? Of course, having honesty and integrity drilled into our psyche, over and over, since Day One, we rolled over and slept until a real emergency surfaced. But I'll tell you, before I knew better, the first couple of experiences were illuminating. When I heard the fire alarm, I got up, opened the door, and went into the hallway. What did I see? Fire? Smoke? No, none of that. I saw scantily-clad civilian girls poking their heads out of doorways too numerous to count. They obviously had integrity battles, too, but slightly different ones. Outside, girls were in various states of undress, surrounded by Marines in various states of undress, all wrapped in blankets, with everyone bitching up a storm. Group grumbling. And promises to kill the alarm-puller. *Proud to serve!*

Dirty Duty

We did have duties similar to fire watch. They were called CDO and JOOD duty. Over the six months of TBS, I was assigned CDO (Company Duty Officer) duty twice and JOOD (Junior Officer of the Day) once.

CDO duty required the presence of two lieutenants. It spanned 24 hours if it fell on a Saturday, Sunday, or holiday. If it fell during the week, it was 12 hours, for it started at the "close of business" and ended the next morning when the training day began. The CDO office was on the same deck as my company in Graves Hall. There was a rack, desk, and phone in the office. The office was a converted BOQ room. Both lieutenants slept in the office, and because duty was really glorified phone-watch, they could sleep at the same time. The attire was the same as fire watch attire: cammies with cover, boots, and web-belt.

JOOD duty required the presence at one of three posts: Mainside (the main part of Quantico), Support Battalion, or O'Bannon Hall. Those lieutenants who were assigned either Mainside or Support Battalion had to draw a pistol from the armory before reporting. (The post at O'Bannon did not authorize wearing a pistol.) With the exception of wearing a pistol and being alone, JOOD duty was indistinguishable from CDO duty. You made rounds and completed the logbook and wore the same uniform. And you watched the phone, like a Maytag Repairman. The hours of duty were the same as CDO.

I should mention the Brig. If a lieutenant was assigned to Mainside, he made rounds through the Quantico Brig. The term "Brig" got its name back in the days when ships were powered by sails. A type of ship back then was called a "Brig." They would anchor it, without sails, in the middle of a harbor. All the military miscreants were housed on it, creating a mini-Devil's Island. In modern times, the Brig is in a building like a county jail. Once a lieutenant enters the Brig, the guard would sound-off, "Attention on deck," and all the military miscreants would come to attention. The lieutenant, feeling somewhat important for the first time at TBS, would glide past them, telling them to carry on. This was the first of many such retorts made by a Marine officer during his career.

Uniforms

The one thing I never understood about my Marine Corps' experience was *why* the Marine Corps did not pay for our uniforms. It was a shocking fact. Service members—Soldiers, Sailors, Marines, Airmen—had to fend for themselves, reach into their piggy banks, and buy camouflage utilities, dress uniforms, and even combat boots. All purchased by people who didn't have a lot of money to begin with. Never mind that all *other* equipment and gear was provided free of charge. Cold weather clothing. Free. Wet weather clothing. Free. Helmets, flak jackets, deuce-gear, weapons, first aid kits. Free. Never mind that. And never mind the difficulty of arguing the distinctions. I guess the DOD needed to save their money to pay for F-117 stealth fighters, B-2 Bombers, and other big-ticket items like $500 wrenches—not for the poor Marine on the ground. *Taxes Being Squandered.*

Not getting uniforms was my first surprise. My next surprise, which was even bigger, was the cost. Before going to TBS in 1991, I was told time and again, from Officer Selection Officers (OSOs) to active-duty Marine officers, that I would spend $2,500 on uniforms. *Twenty Five Hundred Dollars!* I thought they were crazy. No way was I paying that. Absolutely not. But, of course, in the end, I said, "Yes, Sir," and paid the $2,500 for my uniforms, to include a nice Mameluke sword made in Germany, which cost $475.00. Even though we received lieutenants' pay during TBS, $2,500 was still a good chunk of change. There was no legitimate way around paying $2,500.00 for the uniforms. At the time, only two businesses that sold Marine officer uniforms existed at Quantico, and they effectively formed a duopoly (the price difference between them was not stark). The only way around this was to purchase used uniforms from a similarly-sized Marine officer who had no further need for them. But I knew not one. So pay I did. *The Broken System.*

Before we graduated TBS, we had to own and have tailored every Marine Corps' uniform authorized for wear. We were required to purchase four different uniforms: Dress Whites, Dress Blues, Alphas, which are green, and Charlies, which are the everyday working uniform at most Marine Corps' bases (green trousers with brown long- and short-sleeve shirts—three sets required.) We also had to purchase two formal dress covers ("hats," to civilians), white and green, along with white and black shoes. Every uniform had to be fitted, tailored, and inspected by our SPC. That's why we were told *not* to purchase the uniforms before coming to TBS. Our SPC required the whole platoon, at the close of business on a weekday, to wear one type of uniform so he could inspect it. Then, the next week, he required a different uniform, and so on. He told the tailor what needed changing, and the tailor, who came from one of the two shops, disappeared to make the changes. Then he returned for the next inspection, and the one following that one. This ran on for the entire duration of TBS. At the end of TBS, we were inspected, in platoon formation, wearing each uniform. The TBS Staff wanted to ensure that all lieutenants reporting to the FMF had properly fitting uniforms. And that we looked fit and trim.

There was one small consolation to the costly uniform purchase. Most lieutenants didn't have to purchase camouflage utilities at TBS, because they were already purchased at OCS.

MOS Selection

One week before the War, in our 19th week, we were assigned MOSs. When everyone joins the Marine Corps, they have an expectation, a want, a desire to do one particular thing during their tour or, in some cases, their career. Some want nothing but infantry MOSs, others want tanks, and some even want data processing. But whatever the desires, nothing was certain until the 19th week—except for lawyers and student aviators who had "guaranteed" contracts, which promised them either a billet as a judge advocate or naval aviator (provided they passed their respective MOS schools). Consequently, lieutenants with guaranteed contracts had nothing to worry about except the location of their duty station. But all the others had to wait and sweat. And make no mistake, you could look around and see the tension. People's moods changed by the minute. They were acting funny, nervous. Most lieutenants had a whole lot riding on MOS selection. Once they got slotted into an MOS, it would be hard, very hard, to change it. The Marine Corps would spend too many fat dollars training him in his MOS. For the Marine Corps, the incentive to transfer people was small.

MOS selection worked this way. Lieutenants wrote their five most-desired MOS selections on a 3-by-5 card. The company staff took the cards and began the grueling process of assigning MOSs. In the first month of TBS, the company staff posted a list of all available MOSs on the company board (with current lineal standings next to each lieutenant's name). The list was a break-down of how many slots were open for each MOS. The list had (and this is a hypothetical example): Infantry (62 openings), Military Police (2 openings), Tanks (9 openings), and so on. Hence, we were given some idea in the beginning of TBS of our chances in obtaining each MOS. The assignment of MOSs was based on one thing: a lieutenant's lineal standing in the company. But each company staff was afforded discretion on how to interpret lineal standing. For example, they could give all lieutenants in the "top third" their first choice. Or they could give the top 10% of each third (top, middle, bottom) their first choice. Or they could give the top 20% of the "top third" their first choice, and have everyone else get their second, third, fourth or fifth choice. Again, like most things at TBS, it depended on the company staff. Our staff gave the top 10% of each "third" their first choice. When they didn't give a lieutenant his first choice, they called him in and gave options. The whole experience came to a quick conclusion when each platoon gathered in a conference room. Each lieutenant's name was called out, one by one. You'd hear, "Lieutenant Skul, Artillery…Lieutenant Tacelli, Infantry…Lieutenant Thurston, Data Processing…" and so on. As names were being called out, I looked around and saw many lieutenants who were jumping, high-fiving, ooh-rahhing, bursting with excitement, while others looked as if their dog died. It was traumatic.

MOS Selection

PART 2: *THE PACKAGES*

The substantive subjects taught at TBS were called "Packages." These "Packages," to name a few, were Swim Qualification, Land Navigation, Physical Training, Rifle and Pistol Qualifcation, Weapon's Practicals and Live Fires, Patrolling, Platoon Attacks, Defensive Tactics, Offensive Tactics, Artillery Call-For-Fire, Helicopter Operations, Nuclear, Biological, Chemical training, and the longest lasting Package—the Harassment Package—a catch-all package for anything not covered by the other packages.

TBS blended these Packages, including the Harassment Package, where our main worry shifted from week to week. Importantly, these Packages did not proceed sequentially, where one Package was completed each week and you then moved onto the next Package. Rather, each Package stormed into our lives two to three times a week, for a few hours, then stormed out. Each week contained an admixture of Packages, with a Package lasting from one to four months. We had, for example, one or two Land Navigation exercises one week, lasting six hours, along with classes on patrolling, defense tactics, helo-ops, and so on. Then the next week, we did not have any Land Navigation clases; rather, we had Artillery Classes on the fine art of call-for-fire, some PT events, and Field Exercises (FEXs), requiring us to spend from one to four nights in the field. And sometimes we had to go to the field for a tactical evolution or a weapon's live-fire. You never knew. It was always an exciting blend, if you want to characterize it as such. A week's schedule always included classroom instruction (unless it was during the 9-Day War), and it usually contained field exercises and some sleep… if we were lucky. Tired Before Sunrise.

A word about "Remedials." Remedials were extra training sessions for a failed event in one of the Packages. I learned early on that most lieutenants feared Remedials. "The one thing I kept hearing about before reporting to TBS was Remedials, and how bad they were," said 1stLt Craig Wilkerson, a fellow 5th platoon comrade and lawyer. Well, they were *bad. These extra sessions usually took place after the "close of business" on weekdays, or on early Saturday mornings. Remedials were feared not only because of the extra work (you usually had to pass each failed exam* twice*), but because they could, at times, ruin the precious commodity of liberty.*

A Human Sprinkler

Marines are America's amphibious force in readiness. Consequently, the Marine Corps puts considerable emphasis on making sure Marines know how to swim well. At TBS, lieutenants are introduced to "Swim Qualification"—or "Swim Qual," as it is known. Every lieutenant at TBS must attend and pass Swim Qual. It is an arduous package. "For me, Swim Qual was the hardest thing at TBS," said 1stLt Troy Taylor, an Alpha Company graduate. "They almost drown me. I couldn't believe how difficult it was."

Swim Qual comes early. In his "Welcome Aboard Brief," Major Johnson, said Swim Qual was going to commence the very next week, in Week Two. He followed this news with even more good news: "I won't be around much in the next few weeks because I've been tasked with a JAG Manual Investigation into the drowning death of a lieutenant during Swim Qual."[5] A lieutenant had drowned in the pool? Oh my gosh! Now, please understand. I'm not a good swimmer. And that's putting it mildly. So when I heard "drowning" and "Swim Qual" in the same sentence, I thought this was it…I'm going to die and I just got here. Well, as you'll see next, I did not die. But I almost drowned a good friend, who happened to be a terrific swimmer. First, some background.

There were four levels of Swim Qual at TBS (listed from easiest to hardest): S-3, S-2, S-1, WSQ. All lieutenants

---

5   The Marine who died was 2nd Lt. Anthony L. Marchinda, 24, of Troy, Michigan, a city near my hometown of Warren. He died on Saturday, March 16, 1991.

started at S-3 and proceeded up the scale. To graduate from TBS, one had to be at least S-2 qualified. If one wanted to be a pilot, Naval Flight Officer (backseat pilot), infantry officer, or command Armored Attack Vehicles (AAV), one had to be S-1 qualified. No billet in the Marine Corps required a WSQ (Water Survival Qualified) rating; obtaining a WSQ rating simply qualified one for more points, which helped a lieutenant's overall company standing, which helped his chances of obtaining his desired MOS and duty station. And when lineal standings were often decided by thousands of a point, it was wise for some lieutenants to get WSQ. For me, this would not have been wise, as you'll see.

During the second week, my company of 230 was divided into groups of 50-60 lieutenants. Each group was supposed to report to Ramer Hall, the natatorium containing the swimming pool, gym, and weight room. It was located on TBS-proper, about 400 meters from the BOQs. My group of roughly 50 lieutenants reported to Ramer Hall at 0600 hrs on a Tuesday. We wore cammies and boots. But we also brought along other things. The night before, the company staff posted a list of equipment to bring to the pool. It included one pair of boots, two pairs of utilities, socks, and underwear, all of which had to be waterproofed. Waterproofing meant we had to put these items into garbage bags, squish the air out, and then seal the openings with duct tape, making them look shrink-wrapped. We inserted each shrink-wrapped bag into our ALICE packs. This, as we soon found out, prevented us from taking on the properties of a brick.

This first morning, we qualified for S-3 and nothing more. S-3 contained two events. Both events required every lieutenant to wear utilities, combat boots, a Kevlar helmet, deuce-gear, and the stuffed, water-proofed ALICE pack. Everyone also had to carry a rubberized M-16A2. Once we arrived, we were briefed as a group sitting in the bleachers on how to perform the tasks of S-3. It was there, sitting in the bleachers, thinking about the death investigation, smelling the chlorine, feeling the humidity, looking at the glassy surface of the pool, not to mention the tired and stressed faces of my companions, that I began to wonder *why* I had chosen the Marine Corps, and *why* I had signed that USMC contract back on February 22, 1988. Nevertheless, it began...

For S-3's first event, I jumped into the shallow end of the pool, in single-file with other lieutenants, all my equipment on, my rubberized M-16A2 at port-arms. I proceeded to walk across the width of the pool and then back, angling each time to the deep-end. With each crossing, the water kept getting deeper until it was over my head. Fortunately, I knew what to do. I didn't panic. During our briefing I was taught a new and improved swimming technique. In the past, instructors taught lieutenants to swim using a modified breast-stroke—modified because the breaststroke had to be completed with one hand holding the M-16A2. But it really turned out to be a modified mess. Not only was it hard to breast-stroke with an M-16A2 in one hand, the ALICE pack would rise off the lieutenant's back, slide forward into his helmet, and push his face underwater. Some time ago, as I understand it, one lieutenant got tired of blowing bubbles. He turned on his back, put his rifle on his chest, and proceeded to breathe and pump his legs, as if peddling a recumbent bike. The pack, if it was waterproofed properly, was bouyant and kept a lieutenant afloat. As you can imagine, I was glad I was taught this technique. And I was glad I had a buoyant pack. As the water got deeper and deeper, I just laid back, and let my pack do the lifting. I peddled my legs and traveled in the direction I was looking. I peddled two or three times across the deep end, ultimately arriving at the ladder in the corner. This ended the first event. I thought S-3 was pretty easy. But it wasn't over.

The second event for S-3 required the same equipment as the first event. But this time we had to climb a five-meter platform, jump feet first, surface, and bicycle-kick 10 meters toward the far end of the pool, 25 meters away. After 10-meters, we had to remove the ALICE pack, while holding our M-16A2 out to the side, touching neither the bottom nor the sides of the pool, and place our M-16A2 on top the pack and push it to the other side of the pool 15 meters away. For the last 15 meters, we were to use the "frog-style" kicking method such as the one used for the breaststroke. That was it for the second event of S-3. It sounded easy enough.

I lumbered up the stairs with my heavy equipment and stood in line, five meters above the pool. I was dripping and uncomfortable. When it was my turn, I walked to the edge, looked forward (not down), as taught, and took a deep breath. I cusped my chin with my right palm, pinching my nose between my curled index and middle fingers, as taught, holding my M-16A2 with my left hand. I jumped. I surfaced in what felt like a millisecond and bicycle-kicked 10 meters. I stopped, treaded water, took off my pack, and placed my rubberized M-16A2 on top of the pack. Then the fun began. I started kicking my pack and rifle the remaining 15 meters. I kicked and kicked and kicked. But I did not go forward. I went…backwards. Yes, backwards. Ignominiously backwards. A fit and tough-looking swim instructor, dressed in brown swim trunks—they only wore brown swim trunks—was standing on the side of the pool nearest me. *He was making sure I wouldn't drown, right? Because a lieutenant died already, right?* He yelled, "What are you doing?!!" At that point, I could not answer. I was breathing too hard. He didn't care. "Get out of the pool and get back up there!… Do it again!" I climbed out of the pool. I didn't say a word, principally because I couldn't. I was breathing too hard. And with the water dripping down my face over my mouth, and with the power of exhaling, I looked like a human sprinkler. The instructor walked with me to the ladder. Knowing that Plan A didn't work, he created Plan B. He told me to kick my legs in the free-style (scissors-kick) manner instead of the breast-stroke manner (frog-style). I shook my head up and down. I still couldn't talk. On my way up the ladder, I sprayed my oral mist through the steps. At the top was another malevolent swim instructor. He yelled, "JUMP!" Waiting and recovering your breath was not an option. I jumped, surfaced, and started kicking with the free-style scissors kick instead of the breast-style frog kick. It worked. No more reverse gear. But here's the point to all this: the Swim Instructor taught me to adapt and overcome. He saw that the standard kicking style that worked for most everyone—frog-style kicking—did not work for me. So he changed the kicking style to freestyle, and it worked for me. He understood that Marines solve problems by adapting to the situation, changing what needs to be changed, and trying again to see if it works. That's the lesson here. Adapt and adapt until you overcome. For me, at the time, I was just happy I had successfully completed S-3. I was done with Swim Qual for the day. Next was S-2.

The next morning, with the same group of 50 lieutenants, I reported to Ramer Hall for S-2. To say the least, S-2 was more challenging than S-3. It contained a single event and required a water-proofed ALICE pack, helmet, and another Marine, who pretended to be a drowning victim. Before S-2 began, two ALICE packs, one from the saver (me) and one from the victim, were placed on one end of the pool, at its edge. The victim got in the water next to the two packs and waited. I, the purported savior, was at the other end of the pool, 25 meters away. I had to swim to the victim, drag him to where I started, deposit him at the edge of the pool, swim back to the packs, put them in the water and kick them—hopefully forward—to my victim. My victim was 2ndLt. Joshua Skul (pronounced *school*), a U.S. Naval Academy graduate, who was, just the year before, the long snapper on the USNA football team. He was 6-foot-3 and 230 pounds, ran like a gazelle, and swam like a porpoise. I was 5-foot-11, 175 pounds, did not run like a gazelle, and swam like a monkey. After the signal, I swam to Josh, turned him on his back, put him in a headlock, and proceeded to side-stroke him 25 meters to where I began. I did all that and even more. I kneed, kicked, scratched, poked, punched, and slapped him, in addition to dragging him on the bottom of the pool. I will never forget starting out with Josh on his back; his face staring serenely into the rafters, his breathing unperturbed and calm. But after a mere five feet of swimming

The Basic School

and "saving" him, his eyes got big like saucers, his face sank into the water, causing his mouth to spit water like a geyser. Desperation is what his face said. He knew, at that point, he was in the grip of an evil, non-aquatic force, which may kill him. *Was he remembering about Major Johnson's JAG Manual investigation into the lieutenant who drowned? Maybe that lieutenant was a "victim" who drowned at the hands of an evil, non-aquatic force, too.* I got Josh to the other side of the pool. He leapt out of the pool as if it were on fire. He stood at its edge, looking down, hands on knees, dripping water everywhere, trying to breathe and speak at the same time—"you...almost...killed...me...w*hoo!...hoo! hoo!*... everyone...listen...up!...w*hoo hoo hoo!*...NO...ONE... should...volunteer. W*hoo hoo!*...to be...*hoo!*...a vic— tim...*hoo hoo hoo!*...you'll die!" Oblivious to all this—I was exhausted!—I managed to swim back and put the two dead ALICE packs in the water and kicked them to the other end of the pool, where Josh was. And, yes, this time the packs went forward, not backward. We both survived, somehow. Lest it be said, this was an extremely difficult exercise. But it was undoubtedly more difficult for Josh who morphed, in short order, from fake victim to real victim.

S-2 was now completed. For me, being an aspiring Combat Lawyer, Swim Qual was over. Thank my lucky stars! For others who wanted to be pilots, Naval Flight Officers (NFOs), infantry officers, armored officers, and AAVs (Assault Amphibious Vehicles) officers—the fun was just beginning. For them, they had to complete S-1. And by all accounts, S-1 was a significantly tougher test, even for swim-proficient lieutenants.

66

A Human Sprinkler

As far as I could tell from my bleacher seat, where it was much safer and dryer, there were five events for S-1. The first involved swimming four lengths of the pool, using four different strokes: breast, free-style, side, and back. Each of these strokes had to be flawlessly performed. As lieutenants swam, an instructor on the 5-meter platform was watching and screaming out the lane numbers of the lieutenants who had obviously flawed strokes. Those lieutenants could not proceed to Event 2; they had to "enjoy" remedial training, which occurred only on Saturday mornings. Once the strokes were mastered, however, the lieutenant could re-test Event 1 and then proceed to Event 2.

Event 2 involved jumping off the 5-meter platform in utilities and boots only. Before the lieutenant surfaced, he had to reach for the sky with his arms, splash the surface water away, to clear make-believe burning surface oil. Once he surfaced, he had to swim to the far end of the pool using a tiring variation of the breast-stroke, turn, and swim 25 meters to the other end. It was a tiring variation of the breast-stroke, because the upper half of his hands were outside the water; the lower half underwater. This resulted in much water being thrown to the sides—presumably with fake oil fire. This event was extremely tiring, according to every lieutenant who I spoke to. Of course, there was more.

Event 3 involved the creation of flotation devices, all while treading water. The blouse and trousers had to come off to fill them with air. Human air. The pant legs and sleaves were tied in knots to avoid air leakage. They became life preservers, and then were restored to their proper place.

Event 4 involved the saving of another victim. The lieutenant had to jump in the pool, swim 10 meters to the flailing victim, grab him by the collar, swim to the far end of the pool, turn around, and pull the victim across the entire length of the pool.

Event 5, the last event, was a swim of 250 meters, using any stroke. The lieutenant started in the corner of the pool, swam 10 laps, and moved over one lane for each lap completed. Thus, at the end of 250 meters the lieutenant ended-up at the corner across the width of the pool. That finished S-1.

For WSQ, the last and most difficult qualification level, a lieutenant had to tread water for 30 minutes and swim 500 meters. The lieutenant first treaded water for 5 minutes. At the 5-minute mark, he removed his boots, tied his laces together, and hung them around his neck. Then 20 more minutes of treading water. At the 25-minute mark, he untied the boots from around his neck and put them back on for the remaining 5 minutes. Then there was the 500-meter swim. Like the 250-meter swim for S-1, it could be done in any style.

Pump You Up!

The physical training (PT) events at TBS were much more limited than at OCS. Where OCS required daily—and brutal!—organized PT sessions, TBS had none of those. Rather, it had a number of major events, spaced out by weeks and months, that you had to pass. Consequently, staying in shape was our responsibility alone, and it occurred beyond the training schedule. It required self-discipline instead of organized discipline. You had to PT at the ends of very long days—or in between classes during breaks in the training schedule. It was tough to find the time and energy, but it was our responsibility and it had to be done—or you'd fail the events.

To be precise, some organized PT sessions did occur at TBS, although not daily, and they were imposed by the student billet holders. During TBS's six months, I had only one organized platoon run of five miles and three or four other PT sessions (normally two hours), which were run by a student squad leader. There were also times when a student squad-leader would require "individual PT," whereupon some lieutenants conducted Lawyer PT: they read books. Other lieutenants would run...and still others would lift weights in Ramer Hall. *Pump you up!* At other times, lieutenants would play basketball, also in Ramer Hall, play football behind the BOQ, or they would conduct a squad run of 3-5 miles. The company staff would sometimes rise up like a phoenix and have the lieutenants doing platoon or squad runs instead of leaving it to the student billet holders. To make matters worse, some of the squad runs were "Indian

Runs," where you would run in single-file on the road. The last lieutenant in line would break out and sprint to front of the line, settling in and setting the pace. Then the next last lieutenant broke out and sprinted to the front of the line, and on and on, until the run was over.

One thing was clear: the TBS Staff expected us to pass the major PT events with ease. So, again, staying in shape was critically important for each lieutenant, as was practicing each PT event so you could learn to navigate the events and maximize your efficiency, time, and score.

TBS's graded and weighted PT events:

1. The Obstacle Course (O-Course): The Marine Corps has a standard O-Course on its bases. They all have the same obstacles to leap, run, and stumble over, with the same time-limits. A time of one minute or less earned the maximum points. A time of more than two minutes was failing.

2. The Double Running of the O-Course: This event involved running the O-Course twice without stopping. It was designed to blind-side lieutenants who had not mastered the rope climb, which was the last obstacle on the O-Course. If one used excessive upper body strength on the first rope climb, you'd never make the second rope climb. Because of this, it was essential to practice before the graded and weighted event. Lieutenants had to have enough stamina, too, to achieve a good score or to even pass. To get the maximum credit, a time of less than three minutes had to be achieved. To simply avoid failing, a time of less than five minutes would do.

One had to learn to climb the rope by using leg power, not arm power. Lieutenants were taught to reach up and grab the rope, bring the knees up high, pinch the rope between the top of one boot and bottom of the other, stand up, slide the arms and hands up to grab the rope again—and repeat—all the way to the horizontal log on top, which had to be slapped with a hand.

3. The Combat Conditioning Course (CCC): The CCC had five events. The first event was a 3-mile run with helmet, rifle, flak jacket, deuce-gear (with butt-pack), and boots. This event was a ball-buster and was run over the hilly Application Trail. Let me tell you, carrying all this gear up and down the hills of pebble-strewn Application Trail was not fun. Carrying the M-16A2 was the hard part, even though you could carry it any way you wanted, to include slinging it. A 7.5lb M-16A2 felt like a 75lb M-16A2 at the end. This run was timed and had to be completed in under 30 minutes.

The second event was the Fire-and-Maneuver Course, where you run through a course and dry-fire your M-16A2 at targets. This was the easiest event of the CCC. The time-limits were very manageable.

The third event was the Seated Rope Climb, with helmet, M-16A2, flak jacket, and deuce-gear with butt-pack. For this event, we sat at the base of the rope, legs straight out, flat on the ground in front of us, with all our gear on. Once the whistle blew, up the rope we went. And no, we didn't have to remain in a seated position the whole way up. This event had to be completed under 30 seconds and was more difficult than I had expected because of the extra weight.

The fourth event was the Fireman's Carry, where you run 50 meters, pick-up another lieutenant, turn around and run back to the starting line. This event caused some heavy breathing, especially if you had to carry a Marine who was big and heavy. The event had to be completed in under 40 seconds.

The fifth and last event was Push Ups, with helmet, rifle, deuce-gear (with butt-pack), and flak jacket. With this event, one needed to do at least 40 push-ups, with 60 being the maximum.

A failure to pass any one of the five events meant that you had to re-take the *entire* CCC. Most everyone who failed the CCC failed the push-ups. Hardly anyone failed the CCC at OCS. But the difference at TBS—and it was a big one—was the wearing of the flak jacket, and its added weight. One must remember that lieutenants at TBS, although in very good shape, were generally not in as good of shape as they were at OCS. OCS was a physical pressure cooker; TBS was not.

4. The Endurance Course: This was a running course and was the biggest ball-buster at TBS. It was approximately twice as long as the similarly-named course at OCS. It consisted of two 3.2-mile courses strung together, with no pausing or resting. The first course was down Echo Trail, and was over hill and dale with no obstacles, other than a helmet, combat boots, rifle, deuce-gear, and ALICE pack. The second course was through the Stamina Course, which had intermittently placed obstacles to go through or over. There were wood walls with window openings to climb through; criss-crossed, diagonal logs to duck-walk through; carved ruts to low-crawl under barbwire; cargo-nets and horizontally stacked logs to climb over; and approximately six cliffs to ascend and descend using thick, dangling ropes. This run had to be accomplished in less than 90 minutes. No question, this was the toughest 6.4 miles I had ever run. *The Big Suffering.*

5. The PFT: A Marine Corps' Physical Fitness Test (PFT) consisted of three events at the time, with 300 points as the maximum score: (1) a 3-mile run (18 minutes or less was 100 points); (2) pull-ups (20 or more was 100 points); and (3) sit-ups (80 or more in 2 minutes was 100 points).

The substantive distinction between a TBS PFT and an OCS one—other than the minimum criteria differences noted in the next paragraph—was the terrain for the 3-mile run. At OCS, the terrain was flat. Noticeably flat. At TBS, the terrain was hilly. Noticeably hilly. And it psyched-out quite a few TBS lieutenants before they even ran it, because it looked more ominous than it actually was (isn't that a universal truth?). We ran the PFT on MCB-3 (an asphalt road) directly behind the BOQs. We ran 1.5 miles, made a rounded turn, and came back on the other side of the road. The first 1.5 miles were more difficult than the last 1.5, for the last half was relatively downhill. The curious thing was that most everyone's time improved on this course. If you knew how to play it, the hills and the downhill running worked to your benefit.

In the first week, we ran an Inventory PFT. This determined whether we were put on Remedial PT—

something everyone wanted to avoid. The minimum standards for the TBS PFT were not the same minimum standards as the Marine Corps' PFT. The minimum TBS PFT standards were, as follows: a 3-mile run in less than 22:30 minutes; at least 60 out of 80 sit-ups in two minutes; and at least 15 out of 20 pull-ups, which, depending on one's Staff, were strictly scrutinized for full-arm extensions and no rising knees. By chance, if we fell below any of these minimums, then we enjoyed Remedial PT. Remedial PT was orchestrated, at first, by the Company designated PT SPC; later, when he had time to delegate the task, they were orchestrated by the Student Company PT Representative. The sessions normally lasted 1.5 hours and commenced at the "close of business" during the week (if there was sufficient daylight), and on Saturday mornings at 0700. That was the kicker. It ruined an otherwise free Friday night or Saturday morning, for lieutenants couldn't stay out late and drink. More important, however, was the way to get off Remedial PT: a lieutenant had to pass two consecutive PFTs, with the minimum TBS guidelines imposed. A sub-par running of one PFT meant that you were back to square-one and had to run two more PFTs. Not fun. *Testicles Being Slapped.*

The Big Bang Theory

In addition to their physical conditioning, discipline, and "warrior spirit," Marines have another attribute that has been unfailingly consistent throughout their history: marksmanship. Quite simply, wherever Marines are deployed and put into harm's way, they consistently field the best shooters. The Marine Corps places incredible importance on marksmanship. "Every Marine a rifleman" is its credo, regardless of one's MOS. Qualifying on the rifle and pistol ranges was, as you might expect, a long and arduous endeavor.

For two whole weeks, during our 4th and 5th weeks, we spent long days at the rifle and pistol ranges. The worst part of the two weeks was not the range itself but getting there. It was a Royal Pain. Every morning we stepped-off from the BOQ at 0500 HRS in company formation, which meant that wake-up times were around 0330 HRS, and we proceeded to march north, in a column of twos, up the pebble-strewn Application Trail, over hill and dale, with helmet, flak jacket, Deuce-Gear, M-16A2, and ALICE pack loaded with a change clothes. Depending on our student billet holders, and our company staff, our daily marches would be either blisteringly fast or mildly fast—never leisurely. This was one area where a cruel company staff could make our lives miserable. The distance to the range was around 2.5 miles. And since we marched with all our gear and over the hilly Application Trail, sometimes with hard rains turning the trails to slippery, thick mud, the marches could be tortuous. In addition, some platoon commanders made their lieutenants run on the way back, with all their gear. On the way there we never ran, for it was late October and early November, and at 0500 HRS it was pitch black outside. But on the way back, in contrast, there was daylight. Plenty of it. We frequently ran back as platoons and not as a company.

The first day on the rifle range comprised of classes in an outdoor classroom with bleachers. The second and third days were spent "snapping in," as it's called, where we practiced getting into the positions taught to us—sitting, kneeling, standing, and prone—focusing on the target. We dry-fired the M-16A2 while aiming at targets, learning the nuances of the trigger pull itself—the points at which it became harder to pull. You quickly learned there were two or three points in the pulling of the trigger, which were rifle-specific and noticeable "stick points." We learned that you could pull the trigger through these stick points before settling in to fire—which reduced the amount of movement when we finally squeezed the round off. We also learned how to breathe while firing (instead of holding our breath while firing, we were told to uniformly exhale). After three days of this, we went to the actual KD (Known Distance) Course, to fire live rounds and to qualify. We shot live rounds for a few days, practicing our new firing positions and getting a feel for the recoil of the 5.56mm round—and of course for the distances: 200, 300, and 500 yards. No other military unit in the world required its members to shoot at these (or greater) distances. And you didn't get the benefit of a scope at the 500-yard mark. Iron sights were all you had.

The Big Bang Theory

I noticed a peculiar fact while on the range. The Marine Corps used the metric system in most everything they did. Land navigation was in meters, artillery spotting was in meters, calibers were designated in millimeters—but on the rifle and pistol ranges, everything was in American yards under the Imperial System (or under the United States Customary Units). Perhaps it was just tradition being exalted over logic and uniformity and consistency.

At the KD Course, half the company—around 115 lieutenants—became shooters, while the other half went down-range into the "Butts." The Butts, appropriately named, was an area behind an earthen berm where targets were constantly raised and lowered to look for shot holes. There were 50 targets, with two (and sometimes three) lieutenants assigned to each target. The targets were great big things (6'x6'), and had to be carried from the—what else?—Butt House to the target carriages every morning. We worked the targets from a concrete ledge, 3-feet wide and 4-feet high. The ledge was connected to a concrete wall supporting the berm. We did everything from this ledge. Worked on it, ate on it, joked on it, slept on it. Our work in the Butts began once we put the targets in the steel carriages. We raised the carriages and targets above the berm for the shooter—and then waited. And waited some more. We had our backs to the concrete wall, looking in the same direction as the shooter, downrange, waiting for a shot hole to appear in the target. Our view through the frame of the target carriage was another earthen berm— actually, a hillside—approximately 30 yards in front of us. This hillside had botany all over it—grass, weeds, flora and fauna—except for 50 large "ovals" of dirt artistically carved and gouged by thousands of 5.56mm[6] bullets slamming into it. These ovals were important to us. They told us when our shooter shot. When dirt flew, we knew the bullet (projectile) arrived; it was time to yank down the carriage and target and put a circular black or white cardboard "disk" into the shot hole (the disk had a pointed lance in the center to push through the bullet hole in the target). If the bullet hole was in the black, then a large *white* disk was screwed into the shot hole, showing a color contrast. If the bullet hole was in the white, then a large *black* disk was used. After "spotting" the hole, as it was called, we raised the carriage and target, which showed the shooter where his bullet went. The shooter then made corrections to his "dope." Dope? Curiously, the extremely drug-conscious USMC used the word "dope" as a euphemism for front- and rear-sight settings. (Let me tell you, the way the targets looked in the beginning, some Marines must've been on dope.) After five seconds or so, we pulled down the target and removed the colored disk and stuck a square black (or white) "pastie" over the shot-hole. We raised the target again, and the whole process was repeated over and over, from dawn to dusk. A hundred little monkeys reaching, grabbing, pulling, pushing, spotting, pasting, running around, laughing—*heh! heh! heh!*—raising

6 The U.S. military uses the 5.56mm NATO round, which is different in important ways from the comparable .223 caliber Remington round used in the AR-15 and non-military rifles. The first important distinction is chamber pressure: the 5.56 NATO round has a chamber pressure of 58,000 PSI; the .223 Remington round has a pressure of 55,000 PSI. The second important distinction is that the 5.56mm round has a .125" longer throat that not only explains the additional gun powder and pressure, but also explains why the 5.56mm round cannot be used as safely and interchangeably with the .223 Remington round.

and lowering…raising and lowering…all day long. And, by the way, don't think that TBS even thought about providing lance corporals to "work" the Butts for the officers, as they do at some bases; that would have negated the Harassment Package for TBS lieutenants.

Being in the Butts was grueling and monotonous. Not to mention *noisy*. Think about it. You'd hear gunfire from 50 shooters, all firing at different times; you'd hear bullets impacting into the dirt; and you'd hear some lance corporal, in what amounted to an air traffic control tower, in the middle of the Butts, screaming commands over a PA system telling—ordering!—lieutenants in unison to: *"Raise all dog targets in the air!…Reach out and grab ahold!…Yank those dog targets down into the pits!…Spotters, spot those targets!… Hurry, gentlemen!…"* And finally, you'd hear one last thing which initially fooled many lieutenants. Simultaneous with the sound of the bullet impacting the dirt, you'd hear a loud and high-pitched *crack!* Like a Holy snapping of the fingers. Many lieutenants erroneously thought this *crack!* was the bullet penetrating the cardboard target or slamming into the dirt, 10 meters beyond. Wrong. It was a sonic boom…a teeny-tiny one. The 5.56mm bullet was like a tiny airplane bursting through the sound barrier, with a greatly reduced sonic boom, making it sound like a high-pitched *crack!* The Big Bang Theory! Fact is, the 5.56mm rounds easily broke the sound barrier; they have a muzzle velocity of around 3,250 feet per second, whereas the speed of sound at sea level is roughly 760 miles per hour or 1,229 feet per second. As an aside, because of these tiny sonic booms, sound suppressors on muzzles were almost useless for weapons firing projectiles exceeding the speed of sound—in other words, most rifles, as opposed to pistols, would never be effectively silenced because the bullet itself, once outside the weapon, travels at a speed far exceeding the speed of sound and creates a sonic boom separate from the explosion of the bullet in the weapon that can be suppressed or greatly silenced.

I should mention eating on the range. Meals were provided in the Butts by food trucks, sardonically referred to as "Roach Coaches." Some lieutenants would grab some food from the Roach Coaches, but many were content with staying away and eating MREs on the ledge in the Butts.

After sending rounds down-range for four days, and pulling down and pushing up targets, we finally got our chance to be tested. First there was "Pre-Qual Day," which was scored but didn't count as a grade. The next day was "Qual Day," which was graded and weighted. And you had to pass. If you didn't pass, you had to, a week later, march back to the range not with your company, but with another company that was a few weeks behind yours. And you had to qualify all over again. We had about 10 in our company of 230 who had to do this.

There were three levels of qualification, from best to worst: Expert, Sharpshooter, and Marksman, all of which rated different looking badges to wear on uniforms. To get a perfect score, you had to score 250 points. To qualify as an Expert, you had to score 220 points or above (the Expert badge was the nicest looking one; it looked like a crested wreath). To qualify as Sharpshooter, you had to score 210-219 points (the Sharpshooter badge was similar to a German

Iron Cross; we called it the "Nazi Cross"). To qualify as a Marksman, you had to shoot 209 points or less, (this badge was the worst looking of all badges; it was called the "Pizza Box" because it resembled one, with concentric circles etched onto its face.) If you qualified as an Expert, you'd move up a notch on the Marine Respect Ladder, because that's how good Marines—true Marines—were supposed to shoot. And an Expert shooter can save Marine lives. It doesn't get any better than that to a Marine.

We also qualified with the pistol, which, at the time, was the Beretta 92F 9mm semi-automatic. But qualifying with the pistol was not a 2-week evolution. It was four days and was completed during the last week on the rifle range. The pistol range was about 500 meters from the rifle range and portions of the company shuttled back and forth between the two. *The Big Shuffle*. On the pistol range, we had "Pre-Qual" and "Qual-Day" just like the rifle range. However, the pistol range tested our accuracy at 25 yards, 15 yards and 7 feet. And we fired rounds at a much smaller, round bull's-eye target. It also tested our reloading and aiming speed by having us hold the pistol "at the ready" (holding it with both hands at a downward 45-degree angle), raise it up to get sight alignment, fire two shots, dump the magazine, reload, re-aim, and fire two more. It was a great exercise and one that had to be accomplished quickly. In total, there were about five different stages to qualifying. And again, you had to pass.

On our last day at the rifle range, we performed a fire-and-maneuver exercise. It was not graded but it was one hairy event. We lined up as a platoon on the 300-yard firing line with our M-16A2s. Forty-six of us in a straight line, facing the targets, spaced 10 feet apart. We fired a number of shots lying behind sandbags. On command, we ran 100 yards to the 200-yard firing line, where we went down on bended knee and leaned against a post and fired. Then we stood and fired. Then we started walking quickly and online, as a platoon—just like Pickett's charge—toward the targets. We continued walking until a target—any of the 50-odd targets—popped-up. Then we all stopped, fired two shots standing, two shots kneeling. Why the hairiness? The line of Marines was never straight, just like Pickett's Charge. It surged, it wagged, it was disjointed, with some Marines clearly out in front of other Marines. Bullets flying right by them. And worse yet, some Marines actually crossed over into another Marine's lane. After experiencing this, I concluded that surviving TBS was never guaranteed.

Elephant's Trunk

If there's a training event at TBS that induces trepidation, and even fear, in Marine lieutenants, it's the Gas Chamber. There is something unnerving about the Gas Chamber—and chemical agents in general, like tear gas and poison gas. They petrify people. I guess it's because an innocuous *vapor*, sometimes colorless and odorless, can bring men to their knees, put their face in the mud, and choke the snot out of them. But it does. Literally. Perhaps military men fear gas because they can't *attack* it, shoot it, bayonet it, or strangle it; they can only passively *defend* against it with masks.

In the 10th week, we hurled ourselves into the Gas Chamber to experience the wonders of CS gas.[7] The one thing I remember about the chamber was its scheduling problem. It was scheduled just a few hours before the double-running of the Obstacle Course, which was a course requiring much stamina and deep breathing. (The chamber was set for 0730, while the O-Course was set for 1300.) We loved to joke about that. I mean, you came out of the chamber, your lungs seared to half-capacity, your pores clogged with CS, and you were then supposed to run the O-Course not once but *twice* without stopping. *Proud to serve!*

---

7 "CS" is an initialism of the first two letters of the surnames of two American scientists who developed the chemical in 1928: Ben Carson and Roger Staughton. "CS gas" is a misnomer, as the compound "2-chlorobenzalmalononitrile (chemical formula: $C10H5ClN2$)" is a solid at room temperature. It becomes a gas that attacks the mucous membranes when mixed with other substances. See https://en.wikipedia.org/wiki/CS_gas.

On our morning to "experience" the gas, my platoon marched up Application Trail with our M-17 Field Protective Gas Masks. Most of the prior-service lieutenants—read: seasoned—were unshaven and wore old, unstarched cammies, buttoned completely to top. They didn't want any more skin exposed than absolutely necessary. And they didn't want layers of skin shaved off their faces. Think of how aftershave hurts on a recently shaven face. Then think about CS gas compared to aftershave. Of course, before going to the chamber, these insightful techniques were passed on to us neophytes, who quickly put them to use. We didn't shave and we buttoned every button to the top. Leaving the BOQ, my platoon marched 150 meters up Application Trail until we saw a dark green and delapidated trailer on our left. It looked like a military mobile home. When we approached, a handful of Instructors told us to gather in a school-circle for a briefing. The Instructors explained a few things: how to enter the chamber, the exercises to perform once inside, how to exit. It was a short briefing. We broke into two groups. The first group, which I was in, donned and cleared masks and prepared to enter the chamber; the second group stood and watched. After a few minutes of adjusting our masks and getting a proper seal, we began to enter the right door of the chamber in single file. Smoke was billowing out the door. As I walked toward the trailer, my breaths were getting shorter and shorter. I could feel myself starting to sweat. The lenses of my mask were beginning to fog up. I was checking and double-checking the seal against my face. Once inside the trailer, we took positions on each of the two long walls, with backs against the walls. It was dark and smoky—and scary. Someone slammed the door. A round can of CS was flaming on the floor in the center of the trailer; it was the only source of light. Some lieutenants started to cough. My eyes started to burn, although I managed to keep them open. My nose was burning. My throat was tingling, making me cough. The smell was a curious blend of rubber and weird, indescribable smoke. I looked around. Some lieutenants didn't properly seal their mask. Some CS gas was entering their masks. They were coughing and frantically putting their palms over their mask filters and blowing hard to propel the CS gas out the edges of the masks, just like they were trained. Their feet stamped the floor. I thought some were about to run through the wall—a human cut-out!—but none did. Slowly but surely, everyone adapted and gained confidence. On command, we were told to remove our masks and to store them in our cases. This was the scary part. We had to take a huge, and I mean huge, breath and hold it until EVERYONE stowed their masks in their cases. You didn't know how long it was going to take. And if you breathed-in once, you were done. You would start to choke and probably frantically run through the wall. I closed my eyes, took a huge breath, held it, and then took off my mask and stowed it. It was hard stowing the mask, because your eyes were closed, and it was all by Braille. After 30 seconds or so, everyone properly stowed their masks. We were ordered to don and clear our masks. Nobody moved slowly. The speed was just shy of panic. Once our masks were on and properly sealed, we exhaled powerfully while pressing our palms against the masks' filters. Again, this blew the evil CS gas out the edges of the masks. We exhaled powerfully, inhaled, and opened our eyes. Everyone was coughing. Our eyes burned. But we could finally breathe. Next, we were ordered to make like a choo-choo train. We turned right, took a deep breath, removed our masks, placed our right hands on the right shoulders of the Marines in front of us, and marched around the trailer. We made our way to the door—with our eyes clamped shut, of course. But this was a problematic maneuver. This choo-choo-train was longer than we thought. Again, if you didn't take a big enough breath, and you breathed in, you were done. Most everyone took that tiny breath. Some took deep mouthfuls. Maybe 10 feet from the door, no more, I couldn't hold my breath any longer. I took a smidgen of a breath. I inhaled just a little CS gas. Just a wee bit. But that was enough.

As you can imagine, when we exited the trailer, the sight was beyond description. Just ask the second group of lieutenants, who could barely stand because they were laughing so hard. As we exited the trailer, we were pushing and yelling and falling over one another. Now, when I say "yelling," I mean trying to yell. Picture us falling out of that trailer with our eyes clamped shut, our mouths opening and shutting like fish, with no sound being emitted except for choking groans, and our arms straight out to the sides, as if we were flying. (Before going in the chamber, we were told to extend our arms straight out so we wouldn't rub our eyes.) Well, there we were, walking planes. And all planes either crash or land, right? We crashed. We dropped to the ground—not all of us, but many—and walked on all fours, in all directions, opening and shutting our mouths, looking straight down as if the antidote was in the grass. Four-legged lieutenants crawling everywhere. Actually, we weren't crawling everywhere. We were crawling in every direction except toward the trailer. Even blind, we knew that much. Maybe we could smell the trailer. Which brings me to this. With all the groans, the opening and shutting mouths, the clamped eyes…add to this a long and opaque mucous string—a thick pendulous glob, a little white, a little yellow, anchored somewhere in the nose, swinging back and forth, sometimes dragging along the grass. Like an elephant's trunk. Like a slick strand of Silly Putty. When some of us stood, it literally touched the ground. Mine? Mine was only about a foot long. I pinched it off and swung it to the side, like a discarded remnant. The wonders of CS gas.

After a short while, our suffocating moans turned into discernible words. Not very nice words. Swear Words. Gestures toward the miserable green trailer. After we tired of that, we took a seat and became spectators for the second group. We had deep pain in our chests. Our eyes burned. Our noses burned. We blinked frequently. Through the blinking, we watched the second group enter the chamber. And after a few minutes, they exited. They looked just like us, walking planes with mucous elephant trunks swinging left and right, convulsing and choking. It made us laugh (macabre Marine humor). And then our laughs became too painful. We started to cough. We stopped laughing, swore some more at the green mobile home, and went back to the "Q," short for "BOQ," and prepared for the double-running of the O-Course. *The Big Suck.*

The Box Search

The Land Navigation package at TBS was like TBS itself: long, grueling, and stressful. Like swallowing a bone. But some lieutenants enjoyed it. "Land Nav provided me with six hours where I was alone, no patrol orders to give, no AIs yelling at me, just me alone in the Virginia woods looking for my box," said 2ndLt James Burack, another Fox Company student, who later became a Judge Advocate stationed at El Toro Air Station, California. While there was some truth to this, Land Nav was difficult. Most everyone couldn't wait to complete it—especially when the training schedule demanded a full day of Land Nav beginning at 0700 the day *after* a grueling three- or four-day field evolution, when we didn't get in until 0200 the same morning. The entire Land Nav package lasted nearly four months and included both day and night courses.

Like Swim Qual, Land Nav started early. By the second week, we had a couple hours of classroom instruction and were put to work at TA (Training Area)-5, which was approximately 1000 meters from the BOQs. Not-So was our leader for this exercise, and he walked our platoon around pointing out hilltops, ridgelines, fingers, draws, and saddles. He then volunteered a lucky lieutenant to take the platoon to a specific 8-digit grid coordinate on the contour map. Not-So repeated the volunteering process four or five times, and that ended the first session. He was, in contrast to many other things, very serious about land navigation. He told us it would be one of the most important skills we would learn as a Marine. And when you think about it, he was right. Marines rarely operate on well-known terrain; they usually are shipped off to far-away foreign lands that are never before seen. And being able to get to the objective is a necessary predicate to accomplishing any mission.

Approximately a week later, there was "Baby Land Nav." For that exercise, we had to locate five or six boxes using a compass. *The Box Search.* These boxes were ammo boxes affixed to the top of a four-foot post. Stenciled on each box was an 8-digit grid coordinate of its exact position (so you knew if you found the right box). Hence the name Baby Land Nav. Baby Land Nav was graded but not weighted, so it wouldn't affect our grade or lineal standing in the company. But what it *could* affect was Saturdays, a precious commodity in an environment like TBS. If we failed to get 70%, only getting three out of five boxes correct, for instance, then we failed the exercise and were put on Remedial Land Nav, which meant we had to perform the entire exercise again the next Saturday morning at 0630. And we had to pass two consecutive Saturday exercises before being taken off Remedial. So much for celebratory Friday nights. Did I mention signing-in? Another cruel way to fail a Land Nav exercise was failing to sign-in. After finding all our boxes and turning in our card to the AI who was located at a central spot in the field, we had to return to the company office in the BOQ to sign-in the "time" next to our name. If we forgot to do this last thing by going straight to our BOQ rooms—and you'd be surprised how many lieutenants did this—we were put on Remedial Land Nav. Even if we found all our boxes. Proceed to have your Saturday ruined. Once again, the Harassment Package rose like a phoenix. Nobody said TBS was fair. And in this regard, it was cruelly unfair. *Total B.S.* But like other unfair things at TBS, it had to be complied with to graduate.

After Baby Land Nav, we moved to a larger training-area (TA-8) where the grid-coordinates were no longer stenciled on the boxes. Because of this, you were never sure whether the box you found was the correct one. We quickly determined that shooting azimuths from boxes was perilous and had to be avoided. We had to shoot them from other terrain features, first identified on our map.

Four or five navigation exercises followed, with anywhere from 5-8 boxes to find during each one. During these exercises, one or two of the boxes were "aerials." "Aerials" involved looking at an aerial photograph of a piece of land and identifying specific points on the photo and transferring them to the map. The process was called Triangulation. The good thing was, these four or five exercises were graded, but not weighted. We only had to worry about Final Land Nav, the final for the Night Navigation Course, and the potential for "Remedial Land Nav."

We had two night-navigation exercises—one was a practice version with two 400-600 meter legs on TA-5, and the other, which was graded and weighted, had four legs of 1000+ meters each on TA-8 around the BOQs. Both courses provided us with unique (and sometimes dangerous) "opportunities to excel." Did I mention Beaver Dam Run, the sometimes wide and deep stream running through TA-8 and TA-5, which was appropriately named? Well, between the 1st and 2nd legs, and between the 3rd and 4th legs, we had to cross it. At night and in the winter. And some lieutenants exercised poor judgment by picking bad entry points, where they literally had to swim across. With nasty little beavers running back and forth. One lieutenant in our company was bitten by a beaver and had to undergo rabies' shots. *Ticks, Beavers, Snakes.* He said the shots weren't bad anymore. Sure. And another lieutenant, while strolling through the woods at night, looking for his box, walked right into a branch. A blunt stick in the eye. He came close to having a permanent injury. He had to wear an eye-patch for weeks. That's what can happen when you walk through the woods at night. *Thorns, Briars, Stickers.* Yet curiously we were never issued safety goggles, even though we were issued so much other gear, for so many different purposes, and safety was supposedly paramount in training. I guess I could understand not being given Beaver Spray, but no safety goggles? Harassment Package!

One last thing about night navigation. You might wonder how we found our boxes at night when we couldn't see. Simple answer. The staff put dimly-glowing, neon chem-lights on top of each box. Now, these chem-lights weren't like lighthouses guiding us into port. We could only see the chem-lights when we got within 40 feet or so. And sometimes, because of the dense foliage, which was omnipresent in the summer months, we couldn't even see them 10 feet away. But there was a bigger problem: there were *other boxes* in the same area, and they also had chem-lights on them. All the boxes, when they were first planted on posts, were evenly spaced, usually 50 feet apart. So if you came into an area where you could look *left* and see a box

25 feet away, and look *right* and see a box 25 feet away—you were presented with a nice conundrum: Which box do I pick? Well, it depended on your drift when navigating. At the beginning of Land Nav, all lieutenants participated in an exercise to determine drift—i.e., whether they drifted left or right while walking long distances. So this "drift" principle was the way to pick the correct box if you ended up precisely in the *middle* of two boxes (and this applied to all day-time courses, as well). If you had a natural tendency to drift right when you walked long distances, then you picked the box to your left. And if you drifted left, you picked the box on your right. Now, if you ended up closer to one box than another (and you didn't have a drift problem), you took the closest one. Easy as that.

Once we passed the final Night Navigation Course, and completed four or five daytime exercises, we were ready to tackle Final Land Nav. For this exercise, which was truly humbling, we were taken by Cattle Cars—a delightful Marine euphemism for aluminum encased troop transport trucks—to TA-16, an expansive training-area covering more than 30 grid-squares (one grid-square is 1000m x 1000m). We never saw this training-area before. (We didn't perform any FEXs on TA-16 until Final Land Nav was over with.) There were 11 boxes, one of which was an Aerial, to find in eight hours. They trucked us out there—way out there!—at 0700 and dropped us off on the side of the road. We were given a grid-coordinate of the roadside position. And away we went. While the whole training-area was used, not every lieutenant used every acre of it. The whole company used it. All 11 boxes were grouped for each lieutenant in an area covering anywhere from 8-15 grid squares. Still huge, but a far cry from 30 grid squares. Now, that doesn't mean it wasn't difficult or that we weren't tired at the end. Because it *was* difficult, and we were exhausted at the end. Our legs hurt badly. By the time of this event, we became proficient and, most importantly, confident in our navigation skills. A very small percentage of lieutenants failed, maybe two percent.

Acting Like LRRPS

I remember when I first heard we would be patrolling at TBS. It was in a class on platoon tactics. The first thing that ran through my mind was this: how does a *patrol* differ from a *platoon attack*? I was curious. Then I immediately thought of the LRRP (Long-Range Reconnaissance Patrols) that were run in Korea and Vietnam by the Army. (The Army even formed the RECONDO [RECON and commanDO] School that included such patrolling practices.) My mind was filled with romantic notions of patrolling, like the Selous Scouts in Africa, both day and night.

The instructors knew that most of us didn't know the difference between a platoon attack and a patrol. So they formed a FEX called "Zen Patrol." It was the first FEX in the patrolling package. One morning we marched 1.5 miles to a training-area that had possibly the steepest hills in Virginia. It started at 0600 with a Patrol Order read by an AI to 15 of us. One lieutenant was volunteered to be Patrol Leader, and another to be Assistant Patrol Leader. Terrain models were built. Rehearsals conducted. And then, around 1100, the patrol began, but with no formal instruction. And, believe me, the Patrol Leader had no idea what the hell he was doing. He didn't know the formations, didn't know the hand signals, didn't know the places where the patrol should stop. He was like the 14 others: clueless. To help matters, the staff exploded a grenade simulator not 20 feet into the patrol. Right next to the patrol leader. The poor guy was now deaf, wobbly, *and* clueless. (Much snickering and giggling from the staff ensued.) *Tortured By Sadists*. But he finally got us through the exercise. The instructors assumed—probably correctly—that everyone would better understand a patrol if they ran one first, without being told what to do. Another innovative TBS teaching method.

We then returned to Classroom 4, where a sand-table debriefing took place. We learned the ways a patrol differed from a platoon attack: A patrol was usually a lot smaller with an average of 15 Marines, and it was both reconnaissance- and combat-oriented. In contrast, a platoon attack contained around 40 Marines and had no reconnaissance mission; it simply attacked. We also learned that while there was only one *patrol* formation—a close parallel to an "arrowhead," with a point man, two flanks, and a slender long body extending from the point to the rear—a *platoon attack* had a few formations: columnar, echelon, and on-line. And patrols, unlike platoons, performed ambushes with efficiency. One final thing differentiated the two. The well-known and well-despised Patrol Order (for platoons, you used the well-known 5-paragraph order).

Ask any lieutenant who's been through TBS and he'll tell you that Patrol Orders were the worst. They were much more detailed than a 5-paragraph order, and they were, logically, much longer. An average patrol order took an hour to issue orally, and about three days to create (a 5-paragraph order could be completed in about 15 minutes). Even though patrols were considerably smaller than platoons, they required planning on the smallest of details. Every man had a minutely-detailed job to execute. His movements were planned. His positions on halts and ambushes were planned. Everything was planned. Getting tagged as a Patrol Leader was a week-ruining event for a lieutenant. He only wanted to get it over with.

Besides the Zen Patrol, there were five or six more patrols, including a couple of night patrols. These night patrols were fun, especially for the lieutenants designated as left and right flank men (flank men were the eyes and ears of the patrol—its security). They were nearly lost forever, to include me. Before the night patrols, the instructors told the flank men to follow the patrol by sound alone. They told us NOT to maintain eye contact—not even hazy eye contact—with the patrol. That wouldn't serve the purpose, they said. We wouldn't be out far enough to detect the enemy. But traveling exclusively by sound at night was not conducive to staying with the patrol. Especially when you weren't supposed to speak or make loud noises. As soon as I got 10 feet away, I couldn't see the patrol. But don't think I couldn't hear them. I could hear these patrolling-novices—*Tactical Boy Scouts*—crushing everything in their path—twigs, leaves, branches—sounding like an enormous wood-chipper moving through the forest. Even so, I had no clue where they were. At night you can hear everything, from the longest distances. You thought they were 20 feet, when they were actually hundreds of yards away. As loud as these guys were, they could have been miles away. And you couldn't yell or talk (we were ordered not to). It's unsettling to be alone in the inky black wilderness. Can't see. Can't talk. Can't figure out which direction to walk. Can't even figure out where the patrol is (let alone civilization)—all the while getting gouged, poked, and tripped by branches and stumps. In the end, miraculously, I somehow hooked-up with my patrol. I was just happy to *see* my patrol. Like finding a bobbing buoy in the middle of the ocean.

In addition to the night patrols, we performed a couple of ambush patrols and one live-fire ambush. The live-fire ambush was a treat. A student patrol leader led our patrol through the woods to the side of a road. We set-up positions alongside the road, and waited 15, 20, 30 minutes. After lying perfectly still during this time—and I mean perfectly still—silhouette targets hanging from a cable came bouncing down the road toward us. Once they were in the "kill zone," the Patrol Leader detonated a simulated Claymore mine—the signal to unleash our firepower—and everyone emptied their SAWs and M-16A2s into the silhouettes. It was great fun, probably because it was a lot safer—no one running in your path, no bullets flying sideways, no rushing, no hitting the deck, no re-aiming, no stress, no exertion. And no fear, just loud noise and holes in cardboard targets.

In the end, looking back, the patrolling package, which started a couple of months into TBS and finished about one month shy of graduation, was a fun package—unless you were the Patrol Leader. Or a blind and mute flank man.

Samurai Warrior

At OCS during the summer of 1988, we had Drill Evaluation and Drill Competition. With Drill Evaluation, we were handed a card with a list of marching commands on it. We had to march our platoon, a 45-man platoon, to the tune of each command. It was a graded *individual* event, not a platoon event where platoons competed against platoons. With Drill Competition, we competed as a platoon against other platoons, with one overall platoon being the winner. Each platoon picked their best cadence-caller, and the competition was off and running.

At TBS, unlike OCS, we were tested on 3 things: (1) *Marching* a reduced-sized platoon; (2) *Inspecting* a reduced-sized platoon as both a Platoon Commander and Company Commander; and (3) *Sword Drill.*

Marching was just like at OCS, except we only marched a reduced-sized platoon. When I say "reduced-sized," I mean 8-10 Marines. When we hit the parade deck to practice as a company, we were broken into groups of 8-10 (this way, each lieutenant had a turn at commanding more quickly than with a platoon of 45). The hard part of this test was learning the commands, the precise words to say. But even harder was learning exactly *when* to say each command, because there was a "preparatory command" and then a "command of execution." And there is supposed to be a delay between each one. Commands coincided with foot falls. You had to learn what "foot" to commence the command from. There were approximately five 2-hour sessions on the parade deck, for all three events. That was not much time. I found myself, like most others, practicing at night and on the weekends.

Inspecting a reduced-sized platoon was probably the easiest of the three events. It was screaming a few commands from a distance, then striding confidently over to the platoon to inspect each Marine. We stood face to face, looking at each other, looking at nose hairs, yellow teeth, and dirty fingernails. Then we right-faced, walked two-steps, left-faced, and did it to the next Marine. Only one Marine had an M-16A2. For him, we'd grab his M-16A2, twirl it like a cheerleader's baton, pull the bolt back, look skyward through the muzzle, twirl it some more—then hand it back. All in a day's work as a Marine lieutenant. And all to pass the exercise.

Now the last event—Sword Drill—was hard. Handling a sword was far more difficult than I had imagined. And it looked easy when I saw the Marine TV commercial where the Marine officer smartly flicked the sword from his waist to his head, the blade coming to rest near his ear. Yet when I first started playing with that thing, I couldn't understand how the guy on TV didn't slice his ear off or poke out an eye. After a few harrowingly close calls, where I half-expected to see my ear quivering on the ground, I gradually got better and more confident.

The Sergeant Major of TBS, who was the highest-ranking enlisted man at TBS, taught us sword drill. For our company, it was SgtMaj Wilson. And let me tell you, he cut one imposing figure. Many of us got confused when we addressed him either in class or on the parade deck. We usually, and it seemed natural when you looked at him, began by saying "Sir"—whereupon he stopped the lieutenant mid-sentence and launched into a well-rehearsed response: "I'M THE SERGEANT MAJOR. YOU'RE THE SIR… (long p-a-u-s-e)…SIR!" It was a classic. Everyone howled! In fact, many of the lieutenants went around for weeks trying to emulate the way he said it. "I'M THE SERGEANT MAJOR, YOU'RE THE SIR…SIR!"

The Sergeant Major, 100-percent professional Marine, still lost it on occasion. Once his boots hit the parade deck, well, let's just say he went back to another clime and place: the clime and place where he was a Drill Instructor at Parris Island. Where previously he had the patience of a saint, there was none to be found when he hit the parade deck. You should have seen him. He was running around to each "reduced-sized platoon" trying so hard to keep his professionalism. To a lieutenant who just tortured a column movement, he said with tightened lips, "SIR…YOU CAN'T REWRITE THE NAVMAC (the "NAVMC-2691," Navy-Marine Corps Guide for drill movements")…YOU JUST CAN'T DO IT!" And of course lieutenants howled again! And they ran around for weeks trying to emulate the way he said it: "SIR…YOU CAN'T REWRITE THE NAVMAC… YOU JUST CAN'T DO IT!"

Another thing I remember about Sergeant Major Wilson. He told us when you issue commands to a platoon, the platoon's responsiveness is directly proportionate to the force of your command voice. In other words, the louder and more forceful you issue the commands, the better the platoon will respond.

A Good Walk Spoiled

At the start of my TBS class, there were three scheduled conditioning hikes, known as marches: a 10-miler, 17-miler, and a whopping 25-miler. *The Big Shuffle.* But toward the middle of TBS, the 25-miler was dropped from the Program of Instruction. Apparently, in other companies, the 25-mile hike was causing countless injuries, forcing many lieutenants to report to their next command with physical debilitations. So it was dropped, unless the company CO could demonstrate to the TBS CO that his company had trained for marches and had, additionally, shown a record of progress. The CO of TBS had to be convinced that the company had attained an adequate level of training to withstand the rigors of marching 25 miles. Our company only went on a 10-miler and a 17-miler (actually the 17-miler turned out to be an 18.5-miler). And both marches were done at speeds slower than OCS speeds (at OCS, you might as well have called them runs, a 10-mile run and a 17-mile run). The 17-miler was over hard surface roads, which are much easier than traveling on trails loaded with pebbles or, worse yet, mud. But the 10-miler, or at least half of it, was done on the narrow, muddy, rocky, and undulating trail, appropriately called the Washboard Trail, which made it a difficult march, more difficult in some ways than the 17-miler.

Our equipment for the marches was minimal. But that was not always the case with every company at TBS. It again depended on the benevolence or malevolence of the Company Staff. On our marches, we only had to wear a helmet, deuce-gear (with a change of socks in the butt-pack), flak vest, and, of course, our beloved M-16A2. But that was it. No ALICE pack loaded with debris. No loaded butt-pack. And no SMAWs, SAWs, M-203s, M-60E3s, or radios. Nothing but the essentials. In any case, despite the relatively flat roads and trails, marching was never a pleasant, looked-forward-to activity. They were long, painful, and tough—and they caused lieutenants to walk funny for days afterward. The marches were particularly hard on the knees, or more particularly, the backs of the knees, which hurt from overextensions.

Just like OCS, we were advised not to drop from humps. Immediately before our first march—the 10-miler—Not-So assembled us as a platoon. He looked more wrinkled than usual, wearing cammies so faded they appeared to be circa 1945. His collars were curled up again, hiding his bars. Then he began:

> "Awright people, real quick. [He loved to call us "people" and "Devildogs."] Devildogs, we're going on a short 10-mile walk through the park. No one should be a hump drop. Understand? It's too easy. But if you wanna drop, there are three, and only three, ways to be a hump drop:
>
> One, if you're walking on stumps;
>
> Two, if your intestines are coming out your anus; and
>
> Three, if you're wearing a sign around your neck that says, 'I am weak.'

Is that clear, people?

Carry on. *Bark, Bark!*"

The platoon went nuts! No-So did it again! He was burnishing his legend!

Bark, bark? What the hell was that? Marines bark by saying *oohh-rahh!* and *arrgghhhh!* Not-So barks by saying *Bark, Bark!* Bottom line: Not-So could bark any which way he wanted to.

We right-faced as a platoon and waited. We were the last platoon in the company. We waited for the other platoons to start marching, and then fell-in marching in a column of twos. (In the Marine Corps, you always marched in a column of twos on forced marches.) We began the march, and then we did something I had never experienced at OCS. We stopped after 10 minutes. Major Johnson began to walk down the road toward us, telling us the break was for "gear adjustment." Although the idea probably originated long ago, it was a good one and much appreciated. Normally the first break occurred after 50 minutes, because of Marine Corps' policy requiring a 10-minute break every hour. But so many times you have an unbalanced load, something digging into your flesh, a pebble in your boot, a sock that scrunched downward—things you don't notice until you start marching. This break helped cure this early on. We made the appropriate adjustments and were ready to march again. Major Johnson was paying attention to the details and looking out for his men. The mark of a good Marine officer.

We continued the march. The pace was manageable. Once again, as on all the marches, my hands began to swell. Many Marines experienced the same thing and knew they couldn't wear rings while marching. During the march, I found myself getting quiet and going internal. I would think about many things, not paying attention to the actual march or the pain it was causing. While marching, I got into a rhythmic motion, putting one foot in front of the other, not thinking about the purposeful, physical exertions. At OCS, during a 15-mile *night march*, I fell asleep while marching at a blisteringly fast pace. It was only for a second or two, but I fell forward into the candidate in front of me and almost fell down. I even remarked about it right after it happened. My fellow candidates were absolutely uninterested. They were breathing too hard and just wanted to keep up and not fall out. Afterward, I started having doubts about whether I had fallen asleep while marching. Can you sleep while exerting that much energy? My validation came at TBS. Not on the march, but by other lieutenants who said they experienced the same thing on marches. More importantly, I found verification in a book, albeit a novel, which we were assigned to read in the first month of TBS: *The Killer Angels*, the Pulitzer-prize winning Civil War novel by Michael Schaara. In it, the author has a passage about Joshua Chamberlain marching. On page 117, this is what he wrote:

> "Amazing. Chamberlain let his eyes close down to slits, retreating within himself. He had learned that you could sleep on your feet on the long marches. You set your feet to going and after a while they went by themselves

and you sort of turned your attention away and your feet went on walking painlessly, almost without feeling, and gradually you closed down your eyes so that all you could see were the heels of the man in front of you, one heel, other heel, one heel, other heel, and so you moved on dreamily in the heat and the dust, closing your eyes against the sweat, head down and gradually darkening, so you actually slept with the sight of the heels in front of you, one heel, other heel, and often when the man in front of you stopped, you bumped into him."

This passage, more than anything else I've ever read about marching, nails it. It's right on. But on this 10-mile march, because it was basically—by Marine Corps' standards—a leisurely march, I did not come close to falling asleep. But even though the march was mild, we were still hurting for a few days after. For me, I was hurting, as usual, behind my knees from apparent over-extension of the legs. Other than that, nothing. Except the loss of weight, which seemed to drop-off with every march. Before entering the Marine Corps, I thought the one exercise that sheds the most weight was long-distance running. Well, after entering the Marine Corps, I learned that the one exercise that sheds the weight the most quickly is a speedy conditioning hike while wearing combat gear and carrying an M-16A2.

The only problem with our marching program, many lieutenants thought, was the interval of time between them. Our second march was literally months after the first one.

And marching is like all other strenuous activities. You must keep at it to stay in shape. So after our second march, when everyone walked hunched-over for four days, there were more than a few grumbles. But again, if a lieutenant does anything well at TBS, it's grumbling. Whining, complaining, and bitching are taken to new artistic and symphonic levels. Marines bitch with the best of 'em. It's part of being a Marine.

Smash-Mouth

In battle, when the chips are down and no bullets are left, it's time for up-close savagery. It's time to fix bayonets and an unholy gleam. It's time for the Wild Beast to emerge from within. Killing at its finest, up close and personal. Now, admittedly, bayonet fighting isn't an everyday occurrence in battle anymore, but it does still happen. So the Marine Corps teaches it, even to its officers. And they teach it not with real bayonets fixed at the end of M-16A2s, but with long sticks, pugil sticks that weighed 16 lbs, padded at both ends—one end to stab the enemy, the other to butt-stroke him, just like an M-16A2. For this pugil-stick exercise, we wore a football helmet, "horse collar," flak jacket, mouthguard, and an over-sized, external "cup." And we carried the pugil sticks. Now. I know I said these sticks were padded, but let me tell you, "padded" in this sense means "concrete." Getting smashed in the mouth or helmet does not feel good. It feels like getting hit with a cinderblock. Those pugil sticks can knock you out. And they can even do more than that. Marine recruits have been killed through the years during pugil stick competitions. Fortunately, no one was killed or even knocked out in our TBS company. But when you get hit in the head, you expect to see your mouth guard on the ground, perhaps with teeth in it.

There were two sessions of pugil stick fighting at TBS, logically named Pugil I and Pugil II (they should have been named Smash-Mouth I and Smash-Mouth II). Pugil I was similar to the one at OCS. It was a platoon competition, and the matches were man on man. *Mano-a-mano*. Each platoon formed a half-circle around a fighting pit and faced each other. Only two Marines played smash-mouth; the rest played favorites and rooted. However, unlike OCS, we didn't fight against another similarly-sized lieutenant. It was the luck of the draw at TBS. Big against tiny. There were no 2-on-1s, as there were at OCS, but we did fight twice. And believe me, I was gasping for serious air when the referee blew the whistle at the end of three minutes.

The second session, PUGIL II, was a little different. Again, it was platoon against platoon. And it was also by the luck of the draw—big against tiny. In Pugil II, two of us, each dressed for war, stood at the edge of the woods, on a hilltop, about 50 meters apart, looking not at one another, but into the woods, down a trail that seemed to curve toward the bottom. When the whistle blew, we each ran down separate trails that met at the bottom. But the way the trails were made, we didn't know who our opponent was until the last second. Then we clashed, and the smash-mouth began. The first killing blow won. They told us before Pugil II, that if we wanted an assured victory, that we should gain considerable velocity on the run, so that we would clash as hard as possible against our opponent. Like two runaway trucks on a collision course. And let me tell you, that's what happened. Two heavily weighted Marine lieutenants, running full-blast, downhill, screaming like Southern Rebels, hoping to flatten their opponent, and then to butt-stroke him to death. The clash was impressive. It was accompanied by a loud *crack!* and a piercing *ugghhhhh!* with one or two airborne pugil sticks. Then it was frantic activity getting up, finding the pugil sticks, slashing, parrying, butt-stroking. Ultimately, the referee blew the whistle—*hhrrreeeet!*—and two battered lieutenants ran up the trail to daylight, one raising his pugil stick over his head, victorious, the other one running head down, with his stick at port-arms, defeated.

Along with the pugil stick competition, we had training in hand-to-hand combat (or close combat). This training was graded but not weighted—it was simply pass or fail and did not affect our lineal standing in the company. The training for hand-to-hand combat was usually conducted on Landing Zone (LZ)-7 directly behind the BOQs and consisted of one to four "LINEs." A LINE is an acronym for—and I'm not making this up—Linear Involuntary Neural-override Engagement. What? For real? Marine initialisms taken to extremes. Nevertheless, a LINE dealt with different situations, such as an unarmed attacker against an unarmed defender; an attacker armed with a knife against an unarmed defender; and an attacker against a defender both of whom are armed with knives. Each of these LINEs had different techniques for neutralizing the attacker. For instance, in LINE 1 (unarmed attacker against unarmed defender) there were six separate techniques to learn for neutralizing the attacker. LINE 1 was the only LINE that was tested, however. We had to participate and learn the other three lines with no tests.

9-Day War

Most schools have final examinations which test everything a student has learned or failed to learn. TBS was no different, except that the final exam wasn't written, it wasn't graded, and it wasn't even in an indoor classroom. But it was still a final exam, because it was a test of everything we had learned about tactics, such as patrolling, platoon offensive tactics, platoon defensive tactics, land navigation, and even stamina. It was called the 9-Day War and it began in our 20th week. The classroom was outside, everywhere. During the 9-Day War, we covered—by foot, truck, helicopter—so many different training-areas, that I couldn't even begin to tell you where its boundaries were. I can only tell you where it began and where it ended: behind the BOQs in the LZ. The only *good thing* about the 9-Day War was that it was not graded and weighted. Consequently, the stress was limited to enduring the elements of nature, physical fatigue, and hunger (and of course more than a little body odor after not showering for over a week). The war in some ways was a big endurance test. I don't think I'll ever be as tired again (unless I go to a real war). In fact, during the last three days of the war, I didn't know one lieutenant who didn't immediately fall asleep as soon as he stopped marching.

The war was supposed to start on a Tuesday evening of the 20th week, and end the next week, on Thursday morning. On that Tuesday, we were supposed to be carted away on helicopters to a far-away training-area. But the weather was flawed—cloudy and rainy—and so the choppers wouldn't land. They were finicky beasts, and safety was supposedly paramount in training (sure, remember the absence of goggles for night land navigation?). So a decision was taken by the company staff not to "helo-out" and instead leave the next morning on 5-ton trucks. Well, once The Word came down on that Tuesday night, an ungodly roar circulated through the BOQ, which gradually rose to the level of insanity. Stereos were cranked. Motley Crue. Metallica. Van Halen. AC/DC. Even the relatively soft Kansas and REO Speedwagon. You heard it all, and all at once…an artistic cacophony of loud sounds, a blending of rock bands…all swirled together with the screams and high-fives of half-dressed lieutenants running around trying to figure out what to do and where to go. It was as if we all won a private date with a movie star. But all we won was one short night: Everyone had to be back at 0500 the next morning to muster at Motor-T. See what TBS does to you? It makes you appreciate the small things and the smallest of timeframes. Anyway, the next morning, muster we did, and embark we did. Five-ton trucks drove us to a training area, somewhere, in the middle of nowhere. Could have been Vermont, for all I knew.

There were two parts to the 9-Day War: a 4-day pre-war, which was actually 4 FEXs strung together: (1) Attack on a Fortified Position; (2) MOUT, Military Operations in Urban Terrain [translation: house-to-house fighting]; (3) Convoy Operations; and finally (4) Vehicle Ambush Patrol. The second part of the 9-Day War was a 5-day maneuver war.

Now, don't be mistaken, even though the 9-Day War was divided into two parts, we were out in the field for nine straight days with no reprieve from the rain, snow, and sleet.

No warm showers. No toilets. No pizza. But the first four days weren't "tactical." For each of those days, we ended the day at dusk, and didn't dig-in defensively; we simply made campfires, smoked and joked, and slept in a sleeping bag in a hootch (two shelter-halves snapped together). Once we moved into the 5-day maneuver war, and got "tactical," things changed drastically. No more fires. No more tents. No more smoking and joking. It was serious business. And serious exertion. It was true maneuver warfare, where any excess baggage would kill you. So after the fourth day, at the end of the FEXs, we purged our packs of unnecessary items—shelter-halves, extra cammies, extra boots, everything else not necessary—and stuffed them into seabags, which were transported back to the BOQ. The only things left to carry were a sleeping bag and some MREs (Meals, Ready to Eat).

For the 5-day maneuver war, the company was broken in half. One half fought against the other, which equated to two-and-a-half platoons per side. We marched with packs over rough and hilly terrain. We attacked in the rain, snow, and sleet. We marched in the middle of the night to "steal a march." We dug-in defensively every evening. And, believe me, digging-in was the last thing we wanted to do after marching and attacking all day in the hills of Virginia. But dig we did. Two-man fighting holes everywhere. Then we sat and waited. We waited for signs of the enemy attacking. Usually, our company ordered a 25% alert (per fire-team), along with 100% "stand-to's" at two times during the night (at 2330 hrs and 0330 hrs). For the 25% fire-team alert, one of four lieutenants in the fire-team had to remain awake, while the other three slept. Half-hour to one-hour shifts were the norm. Anything longer than one-hour was asking too much. Think about it. Typically, the bubba was dreadfully tired, he couldn't talk to anyone, he couldn't see, and throughout his life he'd been told that sleeping at midnight was a good thing. So one hour was the max for these alerts. These alerts, as you can imagine, were not fun. They caused you to miss sleep when it was needed most. Sure, we got an hour here, an hour there, but it wasn't nearly enough. And did I mention LPs? Listening Posts were established 100-200 meters out from our defensive perimeter and were manned through the night. They were two-hour shifts filled with two lieutenants. Every two hours, two *new* lieutenants had to walk out to the LPs. From where did they originate? Fire-teams, of course. Less sleep for the wicked. *Tired Before Sunrise.*

Add to all this our night marches, to "steal a march." Oh, those were fun, coming off fire-team alerts, 100% stand-to's, and LPs. Imagine starting your day at 0230 with a march into complete blackness, with little or no sleep, and with no possibility of sleep—especially meaningful sleep—until the next night. That wouldn't create a happy face. Nor were there happy faces with us. Not one. It quickly got demoralizing, and we quickly gained an appreciation for an infantryman's life. It's an incredibly difficult job—and not a fun life. And when we thought about that life, an infantryman's life, which was constant, not being a "fun life" didn't even begin to describe it. It was far worse than "not fun." An infantryman, dog-tired to begin with, had to fight for his life under the worst weather conditions imaginable, and then see his friends blown to bits—to see them turned into fine pink mists—or see himself get grievously wounded or maimed. Horrifying and pitiless and thankless comes closer as a description. The 9-Day War showed us this. It was more than a final exam, it was eerily realistic.

BASCOLEX

BASCOLEX is an initialism. But isn't that the oddest initialism you've seen? It was so odd, in fact, that most lieutenants never found out what it meant (BAsic School COmpany Landing EXercise). That was OK, though, because the most important thing about BASCOLEX, as far as we could tell, was this: once it arrived, TBS was almost over. After BASCOLEX, only two weeks were left. BASCOLEX started on a Sunday of the 23rd week and ended the next day. Or at least it was supposed to end the next day. Ours was a little different.

The purpose of BASCOLEX was to introduce lieutenants to amphibious warfare. Around 0530 on that Sunday, everyone in the company mustered at Motor-T and boarded white Marine Corps' buses for a three-hour ride to the Naval Base at Norfolk, Virginia. We took our M-16A2, flak vest, ALICE pack, deuce-gear, and helmet. After a bumpy and uneventful ride, we arrived at Norfolk, and everyone ate a late breakfast, either in the Navy chow hall or at McDonalds. In the early afternoon, we boarded a ship, an LPD (Landing Platform Docking), which contained Assault Amphibious Vehicles (AAVs)—the old LVTP-7s (Landing Vehicle Tracked-Personnel), which function equally well on both land and water. Our ship was the USS *Trenton*. And it was a great big thing, with a pointed bow, and a landing-deck to receive helicopters near the stern. After boarding, we were quickly assigned a rack in the enlisted berthing area. This was where I got a lesson in the fine dynamics of a sailor's life. There were three and sometimes four canvas racks stacked on one another. The bottom rack was literally on the deck. No more than two vertical feet existed between each rack. If you were assigned a rack at eye-level, you stood there and gave it a quizzical look. At first you didn't know how to mount it. You thought, *They're joking, right? How do we get in there? Insertion by fork-lift?* But insert we did. Into the human holster. Once gaining a foot and hand hold, we inserted ourselves. However, we couldn't turn over without hitting the rack above—especially when that top rack-mate was also inserted, pressing his canvas into your sacred area, gravity at its finest, and you were pressing your canvas into the lower rack-mate's sacred area.

Around 1100 hrs the next day, Monday, when the ship was heading due east from Norfolk, and approximately 10-15 miles offshore, we descended the ship's well-deck to board the AAVs. The plan was to board them, be expelled from the ship's belly or stern, and then proceed directly to the beach, where one-third of the company had already established defensive positions. We were to dismount, wade ashore, fight, muster back to the buses, eat evening chow, then head home. Simple as that. Now, that's how it was supposed to happen. Our amphibious landing didn't happen; it was called off.

On Monday, March 16, 1992, the weather in Norfolk was cold and windy. It was around 40 degrees with a nasty, biting wind, which created nasty white caps, which caused the Officer-In-Charge (OIC) to cancel the exercise. We said great, turn this ship around and head back to port—Quantico here we come! Not so fast. When the ship left Norfolk, the Captain of the ship, as it turned out, was on a mission—a

Navy mission, not a Marine TBS mission. So when the
Captain of the ship, a Commander-Captain, was approached
by the Officer In Charge (OIC) of the TBS exercise, who was
a lower ranking Marine Captain, the Commander-Captain
raised his aristocratic chin, and, in his best Thurston Howell
III voice, told the Marine Captain to pound sand. He told
this to Captain Freitus, our OIC. The ship wasn't returning
to port for another four days, the Commander-Captain said.
After hearing this, and not knowing what to do, Captain
Freitus marched down from the Captain's Quarters to talk
with the TBS student staff. Much peripatetic pacing ensued.
He then decided to call the TBS staff ashore. Calls were
made and a solution surfaced: Helicopters. Have helicopters
fly back and forth to ferry the lieutenants to shore. But time
was wasting, and the ship was getting farther and farther
away from port. It was late afternoon. And the choppers
wouldn't make the trip after dusk, and they wouldn't come
at all if the ship was greater than 25 miles from shore. We
were, at that time, about 20 miles offshore. So the choppers
could come. A decision was taken, which later came under
biting criticism, to debark the married lieutenants first, so
they could spend the night with their wives. A better, more
egalitarian decision, was to go uniformly from the 1st platoon
to the 5th, not differentiating on marital status or any other
contrived distinction.

Anyhow, married lieutenants reported to the flight
deck with their huge Kapok life-vest strapped to the front
of them vests ("Kapok" is not an initialism; it is a type of
Bombax tree common to China). They waited. And waited.
The birds, two double-rotored CH-46 Chinooks, circled,

landed, and carried off the married lieutenants. They made what appeared to be three trips—each way was roughly a 15-minute flight—before darkness descended. No more flights, for anyone. By that point, the ship had crossed that line, that invisible 25-mile line, which meant that all single lieutenants were in for the long haul. Once again, the Marine OIC marched up to the Captain's Quarters. And once again, he marched down. More peripatetic pacing and more calls. But the helicopters used for the initial ferry refused to come again. Calls were made to the Marine Corps Air Station at Cherry Point, North Carolina. Surprisingly, there were a couple reserve pilots who were willing to crash through that invisible 25-mile line to rescue 150 "single" landlubbers. But they were going to fly CH-53D Sea Stallions, not the double-rotored CH-46 Chinook. That was not good, as a general matter. The CH-53, regardless of any alpha-numeric upgrade to it, was the only aircraft in the Marine inventory where the pilots were given an extra $6000 per year to fly them. The CH-53 had an annoying habit of unexpectedly dropping out of the sky—the tail rotors would fly off and do their own thing, causing the fuselage and its payload to do their own thing and auger into the ground. Were most of us aware of this? You bet. Every one of us. We were told this bright, shining fact in one of our previously held Marine Aviation classes.

So there we were, 60 miles offshore. We mustered to the flight deck and formed "sticks" of 25 Marines. (Twenty-five Marines is one hell of a payload. CH-46s, which have two rotors, take only 18 Marines. Lift capacity has been the one endearing trait of the CH-53—when it doesn't lose its tail rotor and crash and mangle the payload.)

BASCOLEX

There we were, a "stick" of 25 Marines, standing single-file against the bulkhead on the flight-deck. We were wearing our helmets, flak vests, and big, hard Kapoks; we were holding our ALICE packs and M-16A2s. We waited. It was sunny and cold but one of the bulkheads stopped all the wind for us. The sun was heating our faces while we stared into the blue ocean. Then we heard someone yell, "The birds are one mile out." Sure enough. We saw one helicopter, off the stern, making a long arching turn, coming right for us, off the starboard side. The CH-53 was about to land. Everyone was silent, concentrating only on the *whop whop whop whop whop* of the rotors. It landed about 30 meters away. Hurricane force winds hit us. The Navy flight crew ran out and put chocks behind the wheels. We waited silently for the signal to board, our throats drying by the minute. We were given the signal. We started to walk quickly in single file toward the ramp of the bird. As we got within 10 meters, the co-pilot, leaving his seat and descending the ramp, stopped us and told us to form a school-circle. The roar of the blades was deafening. The co-pilot was dressed in a green flight suit and wore a big, white pilot helmet, with multiple, oversized visors (he looked like a human fly). He said, "I want to give you all a Safety Brief." I could hardly hear him—and I was in the front row with four other lieutenants; the other 20 were gathered in a semi-circle behind me. I looked behind me. None of the other lieutenants were giving him the time of day. They couldn't hear a thing he said. Most, however, were engaged in some sort of conversation, probably figuring out which religion took immediate converts. What followed was an amazing Safety Brief; it was more like a Terror Brief. The co-pilot screamed above the rotors:

"GENTLEMAN, IF WE GO INTO THE DRINK, THERE IS AN ORANGE SURVIVAL SUIT TUCKED UNDER THE BENCH YOU'LL BE SITTING ON. PULL IT ON TO THE BEST OF YOUR ABILITY. HOWEVER, SOME OF YOU WON'T HAVE ONE. THERE'S NOT ENOUGH TO GO AROUND. WE DON'T HAVE 25. BUT IT'S REALLY NOT GOING TO MATTER ANYWAY, BECAUSE THE WATER IS TREMENDOUSLY COLD. YOU WON'T LAST BUT 5 MINUTES. BUT IF WE DO GO IN THE DRINK, DO NOT GO OUT THE REAR RAMP. YOU WILL DIE. **I REPEAT: DO NOT GO OUT THE REAR RAMP! YOU WILL DIE!** GO OUT THE FRONT HATCH OR PULL-OUT THE SIDE WINDOWS. THEY'RE VELCROED AND THEY COME OUT EASY. ANY QUESTIONS? (Silence. And more silence.) OK. WE'RE READY. IT'S GOING TO BE ABOUT A 26-MINUTE FLIGHT. GO AHEAD AND BOARD THE REAR RAMP, FIRST FILLING-UP THE STARBOARD SIDE, THEN THE PORT."

After listening to this motivating speech, or at least most of it, I looked behind me again. More of what seemed to be ecclesiastical discussion. Then I turned to my buddy standing next to me, 2ndLt Anthony Steele, who simultaneously turned to me, and we both yelled through the rotor wash: "OUTSTANDING. JUST OUTSTANDING."

I said it with a smile and in a joking manner. Because you had to laugh. I mean, could you *believe* this guy? We all knew about the crash record of the CH-53s, about the additional $6000 extra pay per year, about our pilots being reserve pilots, and about being way past the 25-mile line of death. And then we get a Safety Brief straight from Hell. Laugh. That was all you could do. *Totally Broken Spirits.*

The co-pilot gave us the signal, so we quickly boarded, throwing our ALICE packs in the center of the bird and putting our M-16A2s between our legs, muzzles down. Pointing the muzzles down was supposed to prevent accidental discharges from rupturing the many exposed hydraulic lines overhead. But these lines obviously had been damaged by one too many accidental discharges: They were leaking all over our packs, which was apparently common for choppers in the fleet. Hmmm. Obviously, the CH-53s had more than just tail-rotor problems. We watched the drops soak into the material, becoming bigger and bigger. We looked around to detect nervousness. None to be found. Just silence. (Which really is a form of nervousness, isn't it?) The bird shook, it shuddered, it took off, straining and creaking into the wind. Later, during the flight, I looked at the pilot's gauges. We were traveling at 900 feet, with an air speed of 110 knots. We reached Norfolk in 25 minutes. We landed, ate dinner, and gladly boarded the buses for Quantico.

Around two weeks later, I was sitting in my BOQ room reading the *Washington Post.* Suddenly I noticed a small article. It was about a CH-46E Chinook Sea Knight helicopter crash 60 miles off the coast of Somalia, with 18 aboard. It crashed on Sunday, March 29, 1992. Four Marines died, 14 survived, with five injured. I must have read that article five times.

PART 3: *THE FUN*

At TBS, trapped between the mundane and grueling was the fun. And believe me, it was considerable. Lieutenants would say, "I can't believe we're getting paid to do this." Consider: rappelling down a 40-foot wooden wall and helicopter skid, live fires with almost every weapon in the Marine Corps' inventory, a trip to the Gettysburg Battlefield to learn about real war and real heroes, Mess Night where great food and liquor were served with distinguished military guests imparting their wisdom, a party with Mary Washington College students, and, of course, graduation itself.

Events were not the only fun we had. Perhaps even more fun were the lieutenants themselves. Most TBS lieutenants were hilariously funny. Maybe it was all the stress and pressure and traumatic times when humor got you through it. Whatever. They were just plain hilarious. I remember standing in many lines for bone-crushing PT events with other lieutenants. What was the line like? Were we lamenting our fate? Were we quiet, contemplating the intense pain our bodies would soon go through? No. We joked, we pushed each other, like bear cubs play-fighting, we created an atmosphere of fun, of positive thinking. Right up to the kick-off point. As the event started, then we put on our War Faces. We acted as professionals. But the humor and camaraderie and the atmosphere created happiness…even when pain was near. I also remember many FEXs when it was 35 degrees (or colder) with freezing rain, and we were soaked to the bone. It was miserable and we were slowly dying of hypothermia, or so we thought. But a well-timed, appropriate joke by one lieutenant knocked that hypothermia on its heels. The shivers stopped. All the negative thoughts about our predicament were quashed, smashed. It was a great lesson—repeated over and over at TBS—and it showed us what humor could do during difficult times.

Rappelling

Rappelling was scheduled in the penultimate week of TBS. One inference to draw from such scheduling was that TBS was so close to being over that maybe—just maybe—we could start having some serious fun. Except for those lieutenants who were truly fearful of heights. One lieutenant, 1stLt Craig Wilkerson, had a phobia of heights and hugged the banister the whole way up the stairs. Once he reached the swaying deck on top, which was 40-feet high, he literally crawled on all fours until he reached the dangling ropes; while crawling, he could only bear to look straight down at the wooden planks. To the amazement of all, perhaps even himself, he actually rappelled one time without being KIA or even WIA. For him to do that, considering his phobia, was one clear example of physical courage. Craig's phobia was so bad, he couldn't even drive over a bridge. He'd ask someone else to drive, while he would lower his passenger seat back and put a towel over his face. Thankfully he was a lawyer, with only metaphorical bridges to cross. We imagined him leading a motorized convoy in war, only to come upon a bridge. He'd stare at the bridge…hyperventilate…then turn around and tell his men, "Go back!"

The rappelling tower was next to the O-Course and Confidence Course. Once we arrived at the tower, the instructors taught us how to make a saddle out of 6-feet of rope. We made our own saddle, which, let me tell you, was one scary thought. I guess it's analogous to packing your own chute for the first time in skydiving. After creating the saddle and belaying some lieutenants to the ground, we stood in line on the stairwell. Ascending the stairwell took some time because only so many lieutenants could be at the top, and the ones at the top moved very slowly, if you know what I mean. We rappelled twice, down two different sides of the tower (two of the four sides were not used). Of the two sides we used, one side had a wooden wall extending to the ground; the other had a make-shift helicopter-skid pad four feet from the top, with no wooden wall extending to the ground, just open space. Our first rappel was down the wooden wall. As I stood on top, dangling my heels over the edge, I eased the rope through my hands until I was almost perpendicular to the wall—with my toes barely adhering to the top edge and my back parallel to the ground. I held this position until the Master on top said GO. When he said GO, my right arm, which was planted firmly in the small of my back, shot out far right, and away I went, my left hand out front, above my head. I bounded five to ten feet—it seemed like 500—before I felt like I was going to hit the ground. So I reeled-in that right arm to the small of my back. Then I crashed into the wood wall with my feet. I kicked off again, threw my right arm out, bounded down, kicked off again, and landed on the ground. I untied myself and climbed up the tower for the second rappel.

The second rappel was supposed to simulate rappelling from a helicopter. I backed-up to the edge again, but this time hopped down four feet to the simulated helicopter skid. Then I eased the rope through my hands, becoming perpendicular to the skid and parallel to the ground. When the Master said, "GO!"…I threw my right arm out, and away I went. A speedy trip to the bottom, with no wall to kick off from.

Live-Fires

There were two types of Live-Fire exercises at TBS. One was a "Weapon's Familiarization" Live-Fire (known as a "Fam Fire"), which was actually multiple live-fires on different days with different weapons. The other was a "Platoon" (or "Patrol") Live-Fire. The first one was structured, mechanical, and safe; the second was free-wheeling, fast-paced, and more than a little scary. But either one prompted most lieutenants to say with considerable excitement: "I can't believe I'm getting paid to do this." This comment was frequently repeated at TBS, especially during live-fires.

For "Weapon's Familiarization," or "Fam Fire," we went to the ranges as a company to fire the SAW (Squad Automatic Weapon), M-203 40mm grenade, M-60E3, SMAW (Shoulder-fired Medium Antitank Weapon [read: bazooka]), and AT-4 (which is a clever way of saying 84, for "84mm"), and to throw a live grenade. Now, when I say "ranges," I don't mean the rifle and pistol ranges with paper targets down-range. No, I mean expansive ranges (usually 1000m x 500m) on far-off training-areas, where charred and rusted hulks of armored vehicles are scattered down-range, anywhere from 700 to 1000 meters away. The SAWs, which fire the 5.56mm round, were fired from the prone position. The M-60E3s, which fire the heavier 7.62mm round, and which recoil a whole lot more, were fired from the prone *and* standing positions. And I don't care how strong you were, when firing the M-60E3 while standing, even with its vertical fore-grip to pull the beast down, the muzzle rose dramatically after one-second of sustained fire.

You couldn't keep it level during sustained automatic fire; it was too powerful.

Each of us fired the M-203 40mm grenade. The "203," as it's known, is an M-16A2 with a grenade launcher (a long tube) fastened under its foregrip, running from the magazine well to the muzzle. To load it, one simply pushed the tube forward toward the muzzle, put a 40mm grenade in the tube by the magazine well, then pulled back the tube—sort of a "reverse" pump shotgun. There was a separate trigger assembly, near the magazine well, for the "203." Once cocked and locked, we were ready to rock. Our target was a beaten and bruised armored car, around 350 meters down-range. We first estimated the distance and set the sights. Then we shouldered it, from a standing position, angling the muzzle upward, and pulled the trigger. *Blooop!* The small 40mm projectile—it looked like a big, fat bullet—went blazing toward the target in a lazy arc, easily seen by the human eye. *Kkunngghh!* It exploded with a flash of white smoke. The armored car, from a distance of 350 meters, looked untouched. The 40mm grenade wasn't powerful enough to blast it apart. That wasn't the case with the next two weapons.

Two lieutenants were picked from the company to fire the SMAW and AT-4 (they were too expensive for every lieutenant to fire one); the rest of us watched in considerable amazement. Let me tell you, when a SMAW was fired, it was incredible. It was the loudest weapon in the Marine Corps' inventory, this side of a 155mm howitzer. There was no *whoosh!* or *blooop!* when the trigger was squeezed. It was a tremendous explosion. When the smoke cleared, you half-expected the lieutenant and his weapon to be missing, vaporized. But he was there, and he stood up shaking his head in disbelief. The AT-4 had the same sort of combustion, but on a much smaller explosive scale. It wasn't nearly as smoky, or deafening, or exciting, but it was more comforting for the poor lieutenant who shouldered it.

We also conducted a "grenade toss." It was on the same day, and range, as the SAW and M-203 live-fire exercises. The grenade toss was a simple exercise. We just pulled the pin, threw it as far as possible, then ducked in a 5-foot-high concrete bunker—a "throwing pit." Then we waited. After four seconds, a huge explosion. The ground shook. Sand, dirt, branches, leaves, and cordite rained down on us. But it was the decibel-level that was truly stunning. It sounded like a small thermonuclear device. Aside from the explosion, the most memorable part of the exercise was the briefing.

The EIs gathered 10 of us in a school-circle by the bunkers. One EI had an inert grenade in his hand, using it as a visual aid. He explained how the grenade worked by taking it apart. He showed us the blasting cap, the spoon, the pin. After he assembled it, and after he showed us how to hold and throw it, he moved on to the real thing. He walked over to a box of live grenades and brought one back, still in its cylindrical, cardboard container. He removed it from the container, and, as you can imagine, held it with care, *much care*. He pulled the pin while squeezing the spoon against the grenade (even though the pin was removed, the spoon, if it was pressed against the body of the grenade, would prevent the fuse from igniting and exploding the grenade).

But suddenly, while he was talking and looking around, he dropped the grenade. Accidently. It slipped out of his hands. He lunged for it but couldn't grab it. The grenade hit the concrete bottom of the bunker, then rolled-out onto the hard dirt, three feet in front of us. All of us looked down, wide-eyed and open-mouthed. The world stopped rotating... All sound stopped...Time stopped. Everything was in slooowwww moootion. We watched the grenade bounce... spin a little...roll this way...that way. The whole world, the universe, the Milky Way, watched the grenade bounce, spin a little, roll this way, that way. The EI who dropped the damned thing, wide-eyed as ever, dove over the bunker wall. There was no discussion from us. No deliberating. No time for pride. We knew what to do...Save our hides! All you saw were assholes and elbows running away from the bunker...up the hill...diving head-first into a trench not 15 feet behind us. We were like 10 flowers, planted upside down. We covered our ears and heads (the parts that weren't buried). And waited. And waited some more. No explosion. Nothing. Nothing but laughter 15 feet away in the concrete bunker. We righted ourselves and peeked over the rim of the trench: A handful of EIs belly laughing like you wouldn't believe. Belly laughing at 10 rookie lieutenants who fell for the old grenade trick. Ten dummies who ran from a dummy grenade. For us, the world started rotating again. All sound commenced. The clocks started ticking. Our eyes and mouths returned to normal size, while our shame and embarrassment grew to new heights. The Marine macabre humor. *Tortured By Sadists.*

The "Platoon Live-Fires" were different from "Weapon's Familiarization." They were more like War

Familiarization, because we really didn't know who was shooting who. But believe me, someone was. And, as a matter of fact, it was other lieutenants. No lie. Bullets flew everywhere. The Platoon Live-Fire was supposed to work like this. A Student Platoon Commander was picked to lead his platoon through the woods to a point where the platoon got on-line (everyone standing side by side), waiting for the signal to attack the objective. The signal was typically commenced by the firing of supporting machine-guns located on the flanks. The assault began, and the line was supposed to advance uniformly, side by side, until the objective was reached. But the line sagged, it surged, it contracted, it did a whole slew of horrifying things. Individual rushes became disjointed. Marines veered into other Marines, who were aiming and firing. After a few rushes, Marines were so tired that the aiming part was no longer occurring. It was just hitting the deck—*Ugggghhhhh!...Hoommph!*—and firing. Ricochets. Marines screaming commands. Tracers flying. Some tracers even flew from the side, perpendicular to the line of attack. Really? Hmmm. Where did those come from? In short, it was one scary event where *anything* could have happened, even survival.

Lest it be said, we had to do more with weapons than just fire them. We had to clean them, too. And to clean them, we had to take them apart. So, of course, the Marine Corps had to inject a little competition and stress into the otherwise boring endeavor of weapon assembly and disassembly. *Tension Building School.* It was called the "Weapon's Practical Exam." It was a graded and weighted event, with time constraints for each weapon. It took place in the 20th week and involved the disassembly and assembly of five weapons: the SAW, Beretta 92F, M-60E3, M-16A2, and M-203. There was also a test on the M-2 Browning .50 caliber machine gun (known as the "Ma Deuce" from the M-2 designator), but that didn't involve disassembly and assembly; it only involved setting "headspace" and "timing"—and fielding questions on what to do in the event of a misfire or jam—known as "Immediate Action." In addition to the disassembly and assembly of the five weapons, there was a "Nomenclature Test." For this, an EI asked a lieutenant to pick out a "firing pin retaining pin" or some other oddly-named part from many others laying on the table. Then an EI asked us how to load and unload, and what to do when there was a jam or misfire. That completed the weapon's practical.

Gettysburg

As a company we visited and toured the Gettysburg Battlefield. It was on a Monday, around the 9th week of TBS. Everyone in the company climbed aboard the white Marine Corps' buses at Motor-T for a three-and-a-half hour ride. We were dressed in civilian attire. The tour lasted the whole day and took us through such notable spots as Devil's Den, Little Round Top, Culp's hill, Cemetery Ridge, and Seminary Ridge. The tour climaxed with a simulated charge—actually a walk— of Pickett's charge. Without exception, we thought this trip was one of the highlights of TBS. But the Marine Corps, as always, had to deflate some of that fun by inserting some pain. *The Big Suffering*. Two or three days before the trip, a list of key military men to the battle was placed on the wall by each platoon's training schedule. The names were grouped by Day 1, Day 2, and Day 3 of the battle. Each lieutenant had to pick one key man—usually a general—on one of those days. Then each of us had to, after touring Gettysburg, research our man and prepare a presentation to the platoon on him, using one poster-board as a visual aid. Even though it was considerable work, most of us appreciated the assignment. Much fun was created in the conference room during the presentations.

My presentation was on Daniel Sickles, a Union General, who, during the second day of battle, for no apparent reason, raced his men forward into a peach orchard, leaving a gaping hole in the Union defensive line. The rest of the Union line watched in amazement, thinking Sickles had gone mad. I had much fun with Sickles, and I caused a near ruckus with my continued reference to "hot shards of metal" raining down on

Sickles and his men by the Confederates. My platoon-mates caught that phrase and took off running with it. They barked, "*Hot Shards Of Metal!…Oooohhh raahhhh!…Arrgggghhhh!*" during the rest of my presentation and everyone else's—and even until the end of TBS. You'd hear some of them, out in the field, around the campfire, in the middle of nowhere, just start screaming "*Hot shards of metal!…Oooohh raaahhhh!*"

Mess Night

In our 18th week, on a Thursday, we wore our Dress Blues and attended Mess Night—a formal dinner in the TBS chow hall, the Hansen Room. Traditionally, Mess Nights were supposed to gather all officers of a particular command to recount war stories. It was always an elegant affair, with white tablecloths, candles, superb food, waiters running here and there, and a head table filled with dignitaries and military legends. We had a Marine legend attend and talk to us: Colonel John Ripley, who won a Navy Cross for rigging the Dong Ha bridge in Vietnam on April 2, 1972 with 500lbs of explosives, halting an advance of 20,000 North Vietnamese Army soldiers. Colonel Ripley's exploits were the subject of a book called, *The Bridge at Dong Ha*. He rigged the explosives by hanging under the bridge, using a hand-over-hand, monkey-bar technique, dangling from I-beams and T-beams, all the while being shot at by the NVA. He "remodeled" the bridge and stopped the NVA from crossing the Cua Viet River. If you were wondering how one might earn a Navy Cross, that'll do it.

Needless to say, our Mess Night was a first-class event and a very enjoyable one at that; it included drinking a mixture of liquors from a chrome commode. A chrome shitter on a table with a big ladle. More than a few of us got tanked on one ladle of that stuff. It was the most potent mixture I'd ever seen…or experienced (except perhaps for a concoction called "Artillery Punch," which was experienced ironically in civilian life many years later). Needless to say, Mess Night was a good time. And it got even better afterward.

Around 2300 hrs, everyone in the company went to a classroom in Heywood Hall to watch video skits put together by each platoon, which required much thought and practice ahead of time. Most were knee-slapping funny. Art at its finest. After the skits, around midnight, we gathered outside for "Carrier Quals." For this event, we unfurled a long plastic sheet down a hillside and threw water and soap on it. We hurled our alcohol-saturated bodies head-first down the watered plastic, to simulate the "controlled crash" of a carrier landing. Did I mention clothes? There were none. And it was the middle of winter at midnight. But no one really noticed. Or seemed to care. Too much alcohol from the chrome shitter. And too much fun. Hilarity and laughing and alcohol whisked the cold away. I should mention Major Johnson. Yes, he was there, too. And yes, he performed his own variation of a carrier landing, except he was equipped with a sombrero, a big fat cigar clenched between his teeth, and a tropical shirt. The Man!

As you can imagine, the next day we felt horrible. So what did The Man do? Did he make us go to class at 0700? Did he care that we were up all night and hung-over beyond belief? You bet he cared. He gave us the entire morning off—and then, in the afternoon, we went to a local theatre and saw a movie called *Cape Fear* with Robert DeNiro. All 230 of us. Now that's leadership.

Mary Washington

After approximately four months of TBS, our company commander, Major Johnson, decided to reward us for completing some tough packages like Swim Qual and the Rifle Range. He told us we were going to have a party in a few weeks—on a Friday night, February 14, 1992, on Valentine's Day. A party? At first we thought, *So what!* A party with the bubbas on Valentine's Day? He said a Friday night. We spend 24/7 with the bubbas. Why did we need to ruin a precious Friday night by spending *more time* with the bubbas, especially on Valentine's Day? But we heard him out. He said the party would be arranged by us. He told us to pick five "party representatives," one from each platoon, to go to Mary Washington College (now known as University of Mary Washington), a co-ed private college, located in Fredericksburg, Virginia, about 15 minutes away. The Reps were to post party bulletins, schmooze, employ any covert or overt military technique learned or even imagined to promote the party. Use Marine initiative and perseverance. Major Johnson said the Reps even could wear the infamous Dress Blues when promoting the party. They could also splash on cologne (a force-multiplier, in Marine parlance). If that wasn't enough, he then unleashed his secret weapon: the Marine Corps would provide transportation, both to and from the party. Shiny, white Marine Corps' buses. Chick ferries. We couldn't believe it. We couldn't believe the Marine Corps would ferry young, nubile college girls (and guys, theoretically) to and from Quantico. I began to think like a lawyer. What about liability? What about lawsuits? But that was Major Johnson's point. He didn't want any drunk-driving incidents. Not on his watch. So everyone said, "Party works for us, bring it on."

Well, the days progressed. The cologned Reps in Dress Blues went to the college, posted flyers, glad-handed, shook hands, told the students (read: women) about the party and shiny, white Marine Corps' buses. But they weren't biting. The Reps returned to base sad, almost morose. They said the students were non-committal, had other parties to go to, were going home for the weekend, were studying for tests… yada yada yada…the standard student replies. The Reps only got a few confirmations. Obviously, more work was needed. So out came the Dress Blues again, with cologne. And back went the Reps. Marine perseverence at its finest. But the Reps got more of the same: a bunch of noncommittal answers.

We were getting depressed. We could tell the turn-out was going to be slim, if not non-existent. And we were *ordered* to go to this party. We had no choice. I could picture it: Two-hundred-and-thirty, Marine officers and no one else present. Great. Well, there was nothing we could do about it. Major Johnson was not going to cancel the party. So the Reps tried one more time. Same responses.

Finally, the day of the party came. No lieutenant was excited or eager for the party to occur. Grumbling was all you heard. Most of us knew this was going to be a bad deal. The opportunity cost for us was fairly considerable. Fridays were precious commodities for members in military training. We only had so many Fridays, and this one was going to be lost forever, tied as it was to a forthcoming bad story. But hey, commiserating and bitching always reached symphonic levels in the Marine Corps. So we prepared for the party, put on our Dress Blue Bravos (blue trousers with long-sleeve brown shirt and brown tie), and tried to fix acceptable grins on faces. We assembled, as a company, in the Officer's Club (O'Club) of O'Bannon Hall, and waited for the buses to arrive. Someone said, "The buses are here." We thought, *Yeah, of course they are—and they're empty.* Then someone said, "Oh man, look at 'em." That hit home. We knew he didn't mean the buses...he meant the payload. Most of us went to the doorway to peek into the parking lot. Some buses were still arriving. They were pulling up and— *Tttttsssssssssss!*—air brakes sounded, the doors flung open, and what we saw next was simply amazing. Downright spine tingling. A stream of beautiful girls descending the steps nonstop onto the asphalt. More than you could count! Every bus was packed! We couldn't believe it. What happened to the ambiguous non-committal responses? *I'm going home for the weekend...I've got another party to go to...I have to study for tests.* What happened to the low turn-out? Who knew? Who cared? We were surrounded and, unlike battle, this was a good thing (although Chesty Puller always said he liked to be surrounded in battle so he could shoot in any direction). The beer and mixed drinks flowed like Las Vegas fountains. Here, there, everywhere...groups of three, four, and five Marines, most of whom were 21 or 22, talking with the same amount of girls, who were 18-21. Small bundles of humans: some inside, some outside in the Tea Garden, a sunken swatch of land between Heywood Hall and the two BOQs. It had nice trees, bushes, benches, and tables...

Wait!…

A *Tea Garden*? Really? In the epicenter of manliness and machismo? In a sea of testosterone? A *Tea Garden* where rough-and-tumble Marines pinched teacups with raised pinkies? Say it ain't so! Well, it *was* a Tea Garden. And I'm not sure how it got that disastrous name—or who named it. Any self-respecting Marine would have named it something like, "Guadalcanal Garden"…or "Combat Bunker"…or "War Patio." Those appellations would have gone down a bit easier. Whatever. We were there, in the Tea Garden, inaptly named as it is, gripping our beers in a manly fashion—no pinkies up!—talking away the night. Before long, the human bundles left the Tea Garden, most going back to the shiny white USMC buses, some going on walks with lieutenants. The party ended a little after midnight, around 0030 HRS, on Saturday morning. All the lieutenants then went back to the BOQs for the night. When we woke up later that morning, it was hang-over time. Not to mention Story Time. Saturday and Sunday were filled with very entertaining and lively stories and reverence for Major Johnson, who pulled off a very good party for his men. This is one small example of creative leadership, looking out for your men, and rewarding them for jobs well done. It also is an example of how the TBS Company Staff (or Platoon Staff) could make, or break, your TBS experience.

Tasting the Nectar

Back when TBS started, I never thought I'd get to graduation. It was too far away. Each day was so packed with activities it seemed like you'd never get through the day, let alone to graduation. But alas, it did come and even though it wasn't a great graduation ceremony, it was better than I had expected, and it was better than the one at OCS. Our graduation was on April 2, 1992, a Thursday, and was in the FBI Academy Auditorium, which seats approximately 2500 people (there was no auditorium even close to that size at TBS). Everyone was dressed in their Blues. The company staff was seated on chairs on the auditorium's stage. A small contingent of the Marine Band was sandwiched between the stage and the first row. There was a guest speaker, Brigadier General Stewart, who was the father of 2ndLt Kevin J. Stewart, a student in my platoon. Each platoon sat in one row, and each was called out by microphone. As a platoon, we stood at once and formed a line on the side of the auditorium to climb the stairs to the stage. On stage, we shook hands with the General and CO of TBS, received diplomas, then strode confidently to the other side of the stage, where we descended stairs, and took our seats again.

It lasted an hour. Everyone then returned to TBS to have "finger food" in the Hawkins Room. Parents met the company staff. Parents met Not-So. He told the parents how we suffered for 25 weeks. After meeting Not-So, my parents and I were standing around talking. And then everyone went home. Disappeared just like that. *Poof.* On to the next school.

Tasting the Nectar

My next school was the Naval Justice School (NJS) in Newport, Rhode Island. NJS was a nine-week program teaching lawyers about the Uniform Code of Military Justice (UCMJ) and military law. I had two weeks to report, so I decided to take some leave and return to Michigan. While all the students were packing their things and saying goodbyes, the TBS staff prepared for the next company coming on board, which was the very next Monday, four days later. Can you imagine? Six long months of training and only four days off. Even the staff experienced the Marine Corps' Harassment Package. *Testicles Being Slapped.*

Proud to serve!

Bark Bark!

ACKNOWLEDGMENTS

No book is self-made. Consequently, many thanks go to those who volunteered their time, effort, and expertise to help improve this book and make it appear in-print. Thanks go to Jim Burack, a truly inspirational friend and fellow Marine officer at The Basic School. He reviewed an initial version of this book...and...lest it be said...nuked it—with blast, radiation, electromagnetic pulse, the whole shebang. I was only left with fragments and a big hole to fill. As a result, I collected the fragments, rearranged them, rebuilt them, sanded them, polished them, filled the hole...and *voila*! Here it is. Thank you again, Jim Burack! Thanks also goes to Cynthia Kruska, a non-Marine friend and former co-worker, who provided exceptional guidance and suggestions, especially when it concerned grammar, sentence structure, and keeping sentences in the same tense, past, present, and future. Other non-Marines helped, too, including my father, Roy Smith, who endured multiple requests to read iterations of the book, and who provided common-sense solutions. Another non-Marine friend, Pat Rombach, provided insightful guidance on various topics throughout the book.

A special thanks also goes to some truly amazing writers who have inspired me through the years. More than anyone else, they led me to write for others and to put pixels on screen—and printed words on paper. If it wasn't for them, and their unmatched talents and creativity, this book would never have been written. So...a doff of the cap to Mitch Albom, Tom Wolfe, Malcolm Gladwell, Ron Chernow,

Truman Capote, and Michael Lewis. Whatever you do, read their books. Their writing might inspire you in ways that you never imagined.

www.ingramcontent.com/pod-product-compliance
Lightning Source LLC
Chambersburg PA
CBHW071644210326
41597CB00017B/2114